A PRIMER IN THEORY CONSTRUCTION

A PRIMER IN THEORY CONSTRUCTION

PAUL DAVIDSON REYNOLDS

THE BOBBS-MERRILL COMPANY, INC.
INDIANAPOLIS AND NEW YORK

 • PRINTED IN THE UNITED STATES OF AMERICA • LIBRARY OF CONGRESS CATALOG CARD NUMBER 70-151610
ISBN 0-672-61196-1 (pbk)
FOURTH PRINTING • DESIGNED BY STARR ATKINSON

DEDICATED TO
JOHN TOM AND BARBARA REYNOLDS,
MY PARENTS

Acknowledgments

The faculty of the Department of Sociology at Stanford University during the years 1965–1968, when I was a student there, is primarily responsible for this book. A major focus of the efforts of Joseph Berger, Bernard P. Cohen, and Morris Zelditch, Jr., was to suggest that it is worthwhile to consider what science should look like before plunging in to develop it. However, although this book shares their general orientation and some of the views expressed may agree with theirs, the author takes full responsibility for the arguments and conclusions in this work.

My greatest debt is to the four classes of students who took my course in social psychology at the University of California, Riverside, during the academic years 1968–1969 and 1969–1970. The impetus for this book was my discovery (1) that bright and interested undergraduates have no conception, right or wrong, about theories or the process of developing them, and (2) that there is no useful text that attempts to approach this material at an introductory level. Willing or not, these students served as the testing ground for the assignments described in the Appendix. The students in the Social Psychology course during 1969–1970, the graduate students who took my Theory Evaluation course during the winter quarter of 1970, and the graduate students in Jerry Kline's 1970 winter-quarter seminar in theory construction at the University of Minnesota School of Journalism all provided valuable comments on the first and second drafts of the book.

Mrs. Clara Dean of Riverside, California, has been an extraordinary typist. The quality and efficiency of her work set a standard that is seldom equaled. My wife, Anne-Marie, and the Bobbs-Merrill editor, Mrs. Maria Scott, have been most helpful in imposing the accepted rules of grammar and spelling on my free-style form.

Preface

This book is brief, but it treats a complex topic with many facets. Undergraduates who have reviewed earlier versions have commented that the book was easier to understand the second time that they read it. It is suggested, therefore, that the reader, particularly if this is his first contact with this topic, plan to cover the book twice. Hopefully, it is short enough so that this will not be a burden.

In an attempt to make certain issues related to the evaluation and construction of theories clear, a number of examples of substantive theories have been presented. The purpose of these examples is not to present "perfect" or "ideal" theories but to clarify issues related to the philosophy of science. For this reason, *substantive theories presented in this book have not been evaluated.* The inclusion or exclusion of a substantive theory should not be considered a reflection of the author's evaluation of that theory. If the reader finds the examples confusing or difficult, they can be omitted without interrupting the continuity of the book.

This book is designed to provide a general introduction to the philosophy of science as it can and should be applied to social and human phenomena. In combination with an introduction to research methods, such as that by Hubert M. Blalock, Jr., *An Introduction to Social Research* (Englewood Cliffs, N.J.: Prentice-Hall, 1970), the reader should be provided with a broad and integrated introduction to theory construction and theory testing in empirically based social science.

Contents

A PRIMER IN THEORY CONSTRUCTION

1. Introduction

A scientific body of knowledge consists of those concepts and statements that scientists consider useful for achieving the purposes of science.[1]

The purpose of this book is to describe the different types of concepts and statements that compose a scientific body of knowledge and to indicate what form they should have to facilitate their adoption by a scientific community. However, the ultimate test of any idea is its utility in achieving the goals of science, and good ideas in clumsy form generally gain wider acceptance than poor ideas in correct form, although it may take longer. *There is no substitute for a good idea.*

The purpose of this chapter is to describe the characteristics of scientific knowledge. In order to explain and justify these characteristics, two issues will have to be discussed first: the purposes science ought to serve and the procedure scientists use in evaluating the usefulness of concepts and statements.

FOR WHAT SHOULD SCIENTIFIC KNOWLEDGE BE USEFUL?

While scientific knowledge is basically a system for description and explanation, not everything can be explained by science.

1. If "scientists" are defined as those individuals who create scientific knowledge, then this statement can lead to problems of circularity—scientists adopt knowledge as science, those that adopt knowledge as science are scientists. However, if scientists are defined in some other fashion, as those with certain types of training or those defined as scientists by society (by whatever criteria), then no circularity occurs.

Such questions as "How does the moon affect the oceans of the earth?", "What changes a person's status in a social system?", or "Under what conditions does 'life' (an organism capable of reproduction) exist?" can be approached with scientific knowledge—and frequently answered. All of these questions are related to how or why certain events occur. Such questions as "Why is there a moon?" or "Why are there societies?" or "Why is there life?" are beyond the capacity of science. These "Why does it exist?" questions are more of a religious or philosophical nature and cannot be resolved with an empirically based science. For this reason, the remainder of this book will *not* treat "why things exist" but will emphasize "why things happen," the major focus of science.

Assuming that scientists have completed the task of building a scientific body of knowledge designed to describe "things" and explain why "events" occur, which is obviously not going to happen very soon, what should such a body of knowledge be useful for? Most people would probably want scientific knowledge to provide:

(1) A method of organizing and categorizing "things," a *typology;*
(2) *Predictions* of future events;
(3) *Explanations* of past events;
(4) A *sense of understanding* about what causes events.

And occasionally mentioned as well is:

(5) The potential for *control* of events.

Each of these will be discussed in turn.

Typologies

Of all these purposes, the first is the easiest to achieve, because any set of concepts can be used to organize and classify. For example, rocks can be classified by color, size, weight, strength, crystalline structure, or any number of other characteristics; individuals can be classified by color, size, weight, strength (physical condition), the nature of their cognitive structure, and so on; and social systems can be classified by size, strength (commitment of members to the system), the form of the internal organization, and so on.

Since there is such a large number of ways to organize and classify phenomena or "things," the problem becomes one of determining which typologies (methods of classifying) are the most useful. This leads to a major issue: What criteria should be used to evaluate the usefulness of typologies? Two of the more obvious criteria are that the application of a typology to phenomena should result in, first, its *exhaustiveness*—of all the "things" being classified, there is no item that cannot be placed in the scheme—and, second, its *mutual exclusiveness*—that there is no ambiguity about where each "thing" is to be placed in the scheme. A third and perhaps more important criterion is that typologies should be consistent with the concepts used in the statements that express the other purposes of science.

Prediction and Explanation

Predicting events that will occur in the future and explaining events that have occurred in the past are, except for a difference in temporal perspective, essentially the same activity as long as scientific statements are abstract. Consider, for example, the following statements:

(i) If the volume of a gas is constant,
then an increase in temperature will be followed
by an increase in pressure.

(ii) If the rate of succession (changes in membership) in an organization is constant,
then an increase in organizational size will be followed by an increase in formalization (of the structure and procedures).

Both of these statements have the same form:

Under certain conditions (constant volume, rate of succession)
a change in one variable (temperature of a gas, organizational size) is followed by a change in another variable (pressure of the gas, formalization of the organization).

Statements of this form compose a scientific body of knowledge and can be used for prediction and explanation of scientific events, using a form of explanation adopted from symbolic logic (other concepts of explanation will be described on the following page).

For example:

If	the volume of a gas is constant,	and the temperature increases,	pressure then the increases.
In situation *Z*	the volume of gas *R* is constant,	and the temperature increased.	
Therefore,	the pressure of gas *R* increased.		

This is a form of explanation adopted from symbolic logic (Hempel and Oppenheim, 1948). Notice that no mention of time (in a historical sense, a particular time in history) enters into the first statement in the explanation. Abstract statements of this sort and, therefore, explanations based on them, are independent of historical time. In other words, these statements can be used to explain past events and to predict future events.

Using an identical logical form, statement ii can be used to explain a change in organizational characteristics.

If	the rate of succession is constant,	and organizational size increases,	then organizational formalization increases.
In situation *Y*	the rate of succession in organization *Q* is constant,	and organizational size increased.	
Therefore,	the formalization of organization *Q* increased.		

Again, this logical deduction is independent of historical time because statement ii is independent of historical time and applies to past and future situations alike.

When a statement is useful for explanation and prediction, the concepts contained in the statement can be used to organize and classify. Gases can be classified according to their volume, temperature, and pressure, and organizations can be classified according to their rate of succession, size, and degree of formalization. Therefore, it seems reasonable to expect that, if a statement can be used to explain or to predict, the concepts contained in that statement can be used to organize and to classify (provide a typology).

Sense of Understanding

A further purpose of scientific knowledge, providing a sense of understanding, is both the most difficult to achieve and the most controversial. It is the assumption of this author, to be followed throughout this book, that a sense of understanding is provided only when the causal mechanisms that link changes in one or more concepts (the independent variables) with changes in other concepts (the dependent variables) have been fully described.[2] If a person feels ambiguous or uncertain about an explanation, it is because some part of the causal linkage has been omitted from the description.

In the previous examples predictions and explanations about gases and organizations were logically derived from statements that are best described as empirical generalizations. Under certain conditions, when scientists have a great deal of confidence in the truth of such statements, they are called laws. However, despite the fact that these examples meet all the requirements of logical explanations, it is difficult to consider them complete, for they do *not* provide a sense of understand-

2. Other criteria for determining the occurrence of a sense of understanding have been suggested. Stinchcombe (1968, p. v) expresses dissatisfaction with dependence on a "logical or formal criterion" for evaluating a theory, but his alternative, understanding the "guts of the phenomenon," is not clearly described. Another view is to consider that a sense of understanding exists when the explanation can be translated into an analogous and familiar process. If the new explanation is similar to some existing idea but only applied to a different phenomenon, then a sense of understanding is considered to exist (Hempel and Oppenheim, 1948, p. 145, describe but do not embrace this argument). Requiring all new ideas to be a version of past ideas would seriously hamper innovation, to say the least. A third criterion is related to model building or the simulation of social or individual processes (see p. 111). A model builder may feel that he has acquired a sense of understanding when there is a close fit between the empirical results and predictions from a model, no matter what the basis for those predictions.

ing. A complete explanation would require several such statements that together provide a description of the causal process.

In relation to the example of the gas at a constant volume subjected to an increase in temperature, the following causal process, based on the conception of a gas as a collection of molecules in constant motion, might be proposed to explain this relationship.

> An increase of temperature increases the kinetic energy of the gas molecules.
>
> The increase of kinetic energy causes an increase in the velocity of the motion of the molecules.
>
> Since the molecules are prevented from traveling further by the vessel of constant volume, they strike the inside surface of the vessel more often. (Since they travel faster, they cover more distance and bounce off the vessel more often.)
>
> As the molecules strike the sides of the vessel more frequently, the pressure on the walls of the vessel increases.

The results of this process are summarized in the empirical relationship: as the temperature increases, the pressure increases. There may be other processes that will "explain" the relationship between a change in temperature and a change in pressure and perhaps this one process could be made more explicit, but nevertheless such descriptions seem to provide a sense of understanding.

In a similar fashion the process that relates an increase in organizational size to an increase in formalization might be described as follows:

> An increase in organizational size is considered to be an increase in the number of organization members.
>
> An increase in the number of members will cause an increase in the variation of training and experience of the members.
>
> As the members vary more in terms of their training and experience, their interpretation of rules and procedures will vary more.
>
> An increase in the variance in interpretation of organizational rules and procedures will cause a decrease of coordination in organizational activities.
>
> A decrease in coordination of organizational activities causes a decrease in organizational performance.

A decrease in performance disturbs organizational administrators.

Organizational administrators attribute poor performance to a decrease of coordination.

Organizational administrators attribute a decrease of coordination to ambiguous rules and procedures.

To reduce the ambiguity of organizational rules, the organizational administrators increase the number of rules and make the rules more detailed and specific.

An increase in the number of rules and their specificity is generally considered an indicator of an increase in formalization.

Again, the results of this process are summarized by the statement: as size increases, formalization increases. Again, it is likely that other processes may link changes in size and the degree of formalization.

Two ways of explaining events, derivation from a scientific statement (or law) and description of a causal mechanism, have been presented. Both succeed in explaining, but it is clear that the second procedure, the description of the causal mechanism, provides a sense of understanding that is absent from the first procedure, the derivation from a scientific statement.

If the description of a causal process is properly formulated, it can provide explanations and predictions, in the sense of logical derivations, as well as provide concepts for organizing and classifying phenomena of interest. If the development of descriptions of causal processes is considered the most important purpose of scientific activity, then any typology not consistent with these descriptions will only yield confusion rather than clarity.

Control

If the ability to control events is taken literally as a desirable characteristic of scientific knowledge, then much of the current subject matter of science would be excluded. For instance, both astronomy and geology are considered to be sophisticated sciences in which there exist useful typologies as well as statements that explain and predict while providing a sense of understanding. But to expect astronomers to control such events in the solar system as eclipses or geologists to control such events within the earth as earthquakes is unreasonable since such control is clearly beyond their present capabilities.

The issue is one of making a distinction between understanding how certain variables affect one another and being able to change the variables. In order to control events in a predictable fashion it is necessary to meet both conditions. There is no reason to believe that some social phenomena, such as characteristics of a status structure (e.g., social mobility) or an economic system (e.g., inflation), won't be as hard to control as eclipses or earthquakes, even though social scientists may be able to provide typologies, explanations, predictions, and a sense of understanding with their theories. For the remainder of this book, control will not be treated as a necessary criterion for accepting knowledge as scientific. However, it will be assumed that, if a theory related to a particular phenomenon is scientifically useful, then scientists and "men of action" can examine their ability to influence the variables that will affect the events they wish to control.

THEORY

What is a theory? Thus far there has been little mention of scientific theory, partly because there is more than one concept of what a theory is. Although some of these concepts of scientific theory will be discussed in a later chapter (see Chap. 5), some comments can be made here. Two conceptions of scientific theory currently dominate. One, the conception of scientific knowledge as a set of well-supported empirical generalizations or "laws," can be referred to as the "set-of-laws" form of theory. The other, the conception of scientific theory as an interrelated set of definitions, axioms, and propositions (that are derived from the axioms), is borrowed from mathematical conceptions of theory and is called the "axiomatic" form of theory. Although both the set-of-laws and the axiomatic conceptions of theory can be used to logically derive explanations, neither provides a sense of understanding in the form the theory is usually given.

If a sense of understanding is provided when a description of a causal process is presented, then it seems reasonable to consider a third conception of theory, as a set of descriptions of causal processes; this can be considered the "causal process" form of theory. Frequently, a set of statements in axiomatic form may be reorganized into a description of a causal process, but this is not always possible. Unfortunately, this is seldom the case with the statements in the set-of-laws form of theory.

The word "theory" is frequently used to refer to a number of other types of formulations, usually abstract, including (1) vague conceptualizations or descriptions of events or things, (2) prescriptions about what are desirable social behaviors or arrangements, or (3) any untested hypothesis or idea. To refer to any set of abstract concepts used to describe a phenomenon as a theory is an inappropriate use of the word if *only* a set of concepts is presented (e.g., the characteristics of a bureaucratic organization). At best, these concepts can only provide a typology but fail to achieve the other goals of scientific knowledge. More will be said about this later. Much of past social and political theory describes the form of an ideal social system, with no relationship to the actual types of behaviors or arrangements that develop among the members of social systems. In this use of the term "theory," to refer to any untested hypothesis or idea, its meaning is the same as when it is said that any idea is "theory" until it is supported by empirical data, whereupon it becomes "fact" or "reality."

This and related uses of the word "theory"—be they conceptualizations, prescriptions for behavior, or untested ideas—will be avoided in this book. Unless otherwise indicated, the use of the term "theory" will refer to abstract statements that are considered part of scientific knowledge in either the set-of-laws, the axiomatic, or the causal process forms.

HOW DOES A CONCEPT OR STATEMENT BECOME PART OF A SCIENTIFIC BODY OF KNOWLEDGE?

In general, the degree of acceptance of an idea as part of scientific knowledge increases (1) as each individual scientist becomes more confident that the idea is useful for the goals of science, and (2) as the number of scientists that consider the idea useful for the goals of science increases. Notice that ideas are not accepted or rejected, either part of science or not part of science, but vary in *degree* of acceptance. Although this is generally the most appropriate way to view the process of accepting scientific ideas, occasionally confidence in one widely held theory will be so high that it is considered *the* truth, and anything that contradicts *the* truth is considered false. The rigid classification of ideas as true or false does not encourage the development or acceptance of new ways of thinking about phenomena.

Two factors are important in affecting an individual scientist's attitude toward a concept or statement:

(1) The scientist's confidence that he understands the meaning of the concept or statement;
(2) The scientist's confidence that the concept or statement is useful for achieving the purposes of science.

The second factor is generally dependent on the correspondence between the idea and the results of empirical research. The major importance of empirical research is in its effects on the degree of confidence that scientists have with respect to some aspects of scientific knowledge. This fact places one important restriction on statements or concepts proposed for addition to a scientific body of knowledge: there must be some way they can be compared to the results of objective empirical research.

If an idea cannot be compared with the results of empirical research, there is no way for other scientists to determine for themselves if the idea is useful for the goals of science. An untestable idea is, then, one scientist's view of the phenomenon, not knowledge that can be shared by all scientists. As long as scientific knowledge is that knowledge accepted by a group of individuals, the view of *one* individual cannot be considered scientific knowledge.

Of more importance to the present discussion is the consideration of how scientists determine whether or not they understand the meaning of a concept or statement. Like any individual in a situation of uncertainty, they check their interpretations by comparing it with the interpretations of someone else, another scientist. If there is substantial disagreement among scientists, then no one scientist can be sure that he has correctly understood the meaning of a concept or statement.

The importance of having agreement about the meaning of a scientific concept or statement can best be understood if the alternative is considered. First, if there were no shared agreement on meaning, then scientific knowledge could not be transmitted from one generation of scientists to the next. Each scientist would then have to build a body of scientific knowledge from the same starting point—complete ignorance. It would be impossible to build a very significant or useful body of scientific knowledge under these conditions.

Second, if scientific knowledge is considered to be that knowledge that scientists agree is useful for achieving the goals of science, then scientific knowledge is impossible unless there is agreement about the meaning of a concept or statement. For if there is no agreement about the meaning of a concept or statement, how can there be agreement about its usefulness in organizing, explaining, predicting, or providing a sense of understanding? Under such conditions the knowledge developed by each scientist would be indistinguishable from everyday knowledge about events in general. It would become part of the scientist's personal philosophy about the world and no different from such knowledge held by a nonscientist.

In summary, if scientific knowledge is that knowledge that scientists agree is useful for achieving the goals of science, there must be agreement on the meaning of the statements and concepts that express scientific knowledge, and it must be possible for any scientist to compare some aspect of his theory with empirical research.

DESIRABLE CHARACTERISTICS OF SCIENTIFIC KNOWLEDGE

The desirable characteristics of scientific knowledge are:

(1) Abstractness (independence of time and space);
(2) Intersubjectivity (agreement about meaning among relevant scientists);

(3) Empirical relevance (can be compared to empirical findings).

This section will describe (1) why these characteristics are desirable for achieving the goals of science and (2) how they facilitate the adoption of concepts and statements as scientific knowledge.

Abstractness

In its simplest form, abstractness means that a concept is independent of a specific time or place. In other words, a concept is not related to any unique temporal (historical time) or spatial (location) setting. Why is it so important that concepts used in science be abstract? There are two reasons, one related to the purposes that scientific knowledge should serve and the other a matter of efficiency in developing a scientific body of knowledge.

It is considered appropriate for scientific knowledge to make predictions about the future. However, if the concepts in the statements used to make predictions are not independent of historical time, they must be unique to a specific time. If we assume that the most important basis for confidence in scientific statements is their correspondence with empirical findings (the results of research), then any statement unique to a temporal setting must be unique to a past temporal setting, because the research must have been conducted in the past. If the statement is unique to a time in the past, then it cannot be applied to situations in the future. In sum, any scientific statement supported by research that is specific to a temporal setting must be specific to the past and cannot be used to make predictions about the future. Quite simply, such statements are not useful for achieving all the goals of science (predictions about the future).

The second reason for requiring abstractness is efficiency. If a scientific concept or statement is developed that is specific to a particular spatial setting, a unique location in the universe, then it cannot be used for prediction and explanation at any other location. If this procedure were followed, each different location (or culture) would require a unique body of scientific knowledge. To say the least, this is an inefficient procedure, and the task of science is difficult enough without this additional complication.

However, there are often events of enough importance, even though they are unique to a particular time and location, that considerable effort is expended to explain their occurrence. Particular aircraft accidents, certain geological phenomena, and many historical events (such as the results of a particular election) fall into this category. In each case the event is specific to a particular location and place in historical time: Why did two planes collide over New York City on a certain date? Why is the Grand Canyon now in Arizona? Why did Nixon get more votes than Humphrey in the 1968 presidential election in the United States?

Such explanations, unique to one specific event, can be called *historical* explanations. The most complete and widely accepted historical explanations utilize general scientific knowledge in explaining the unique event of interest. If the use of these general principles is made explicit, it may be that the successful application to the unique event increases confidence in the general principles as useful scientific knowledge. However, many explanations of unique human events tend to neglect the general principles or fail to make them explicit and focus on the specific aspects of the event of interest. Not only does this distract from the use of general principles in explaining human events, it prevents such explanations from contributing to the development of a scientific body of knowledge, since the general principles are never made explicit. (See Popper, 1957, for another discussion of historical explanations and their deficiencies.)

Intersubjectivity (Meaning)

"Intersubjectivity" means shared agreement among relevant individuals with respect to (1) the events or phenomena encompassed by a concept, and (2) the relationship between concepts specified by one or more statements.

The notion of shared agreement with respect to the meaning of a concept is relatively easy to understand. If a scientist uses a term, such as "tree," "mass," or "attitude," and if his audience shares his definition (of "tree," "mass," or "attitude"), then there is intersubjective agreement with respect to the concept. This is most likely to occur when the scientist (1) attempts to be as explicit as possible in defining new concepts, and (2) makes sure that there is shared agreement on any terms used in defining a new concept.

Intersubjectivity (Logical Rigor)

The concept of intersubjectivity with respect to relationships between concepts is more complex. Any statement, at a minimum, describes a relationship between two concepts. Assuming that there is shared agreement about the meaning of the concepts, it is appropriate to consider the conditions under which there is shared agreement about the relationships specified in a statement.

If only one statement is under consideration, achieving agreement about the nature of the relationships is not a major problem. However, if there are a number of statements, which is usually the case, different combinations of statements can be used to develop a variety of predictions and explanations. The complexity of making predictions from sets of statements can cause disagreement among scientists unless there is some agreement as to how statements should be combined for predictions and explanations.

The solution to this problem is to have a logical system, independent of the substantive content, that can be used to specify the relation expressed in a single statement as well as the implications of combinations of statements. There should be shared agreement about the predictions made within this logical system independent of the content of the theory. In other words, the logical system could be used with different theories treating different phenomena.

A scientist has two alternatives for acquiring a logical system. First, he can develop a logical system for use with his theory and then present this system for the inspection of other scientists, independent of the substantive content of the theory.

However, this is a major undertaking, and few scientists have been willing, or able, to develop their own logical systems. Second, the scientist can use a logical system that has been developed by others and is already shared by the relevant scientific audience. Fortunately, there is a rather large set of logical systems available for borrowing. Mathematics (essentially a set of logical systems), symbolic logic, and computer languages can be employed. Many of the logical systems in mathematics have been developed for the specific purpose of providing scientists with logical systems suitable for particular substantive theories, generally in the physical sciences.

The reason for requiring intersubjective agreement on the relationships within and between statements can be stated briefly. If scientists cannot agree on the predictions derived from combinations of statements, then there can be no agreement as to the usefulness of the statements for predicting or explaining phenomena. If scientists cannot agree on the usefulness of the statements for achieving the goals of science, the statements cannot be accepted as part of a scientific body of knowledge.

The requirement for shared agreement about the relationships among statements is usually referred to as the need for "rigor," which is generally taken to mean "logical rigor."

Empirical Relevance

The possibility of comparing some aspect of a scientific statement, a prediction or an explanation, with objective empirical research is what is meant by the criterion of empirical relevance. In order to understand the importance of this criterion we must consider a difference between perceiving an event and explaining it. Perceiving an event is having sensory experience of a particular state of nature. Explaining why one event is associated with another, or what causes an event, is the basic purpose of a theory. Individual perception is a sensitive and delicate process, and often many subtle and unconscious factors may affect what an individual thinks he has perceived. If an individual has proposed a theory and the only source of supporting evidence is his perception of the phenomenon of interest, then it is not clear whether or not his desire for the theory to "work" affected his perception of the phenomenon.

For this reason, it is desirable that any scientist be able to examine the correspondence between a particular theory and objective empirical data. The important factor is that the potential for such a check be available, for very seldom is it initiated. But if a theory and the empirical evidence that supports it is presented in the appropriate fashion, in detail, other scientists feel that they can verify the results for themselves, and this increases their confidence in the usefulness of the theory.

Again it is fruitful to consider the alternative. If a theory cannot be compared with objective research by another scientist, then it becomes the private philosophy of the originator and hence cannot become part of a shared body of knowledge —part of science.

SUMMARY AND CONCLUSION

The attributes that appear to facilitate the adoption of concepts and statements into the scientific body of knowledge are:

(1) Abstractness, independence of time and space;

(2) Intersubjectivity

  (a) Explicitness—description in necessary detail and with terms selected to insure that the audience agrees on the meaning of the concepts;

  (b) Rigorousness (logical rigor)—use of logical systems that are shared and accepted by the relevant scientists to insure agreement on the predictions and explanations of the theory;

(3) Empirical relevance—the possibility should always exist that other scientists can evaluate the correspondence between the theory and the results of empirical research.

The final test of any concept or statement is whether or not it is adopted by other scientists as useful for the goals of science; the above list only describes those characteristics that facilitate this process if the basic idea is useful. If the idea is not useful, then correct form will be of little use in facilitating its adoption.

In this introductory chapter, consideration of the goals of science (providing typologies, explanations, predictions, and a sense of understanding) and consideration of the process of accepting ideas into the body of scientific knowledge have been examined in order to suggest some desirable characteristics for scientific knowledge (abstractness, intersubjectivity, and empirical relevance). In the chapters that follow theory construction is approached from the other direction, from the view of a person planning to develop a theory. Chapters 2, 3, 4, and 5 explain how theories are described: in Chapter 2 the basic idea or conceptualization of the phenomenon is discussed; in Chapter 3, the presentation of concepts; in Chapter 4, the development of statements (describing the existence of or relations between concepts); and in Chapter 5, the organization of sets of statements into theories. The remaining chapters emphasize broader issues: Chapter 6 focuses on testing theories; Chapter 7 discusses strategies for developing theories and contains a few comments on the circumstances under which individuals seem to generate "new ideas," the most important activity in science; and Chapter 8 concludes with a discussion of the possibilities for developing a science of social and individual phenomena.

2. The Idea

Before a scientist begins to describe his "new theory," by developing definitions, statements, and the interrelations between the statements (or theories), he usually has a conceptualization—an orientation toward or perspective on the phenomenon, and this forms the basis for his written or formal theory. In other words, after a new idea occurs to the scientist, he attempts to describe this idea to others using explicit definitions and statements. At times this new idea is more than just a different way of describing the same data; it may include a unique "world view" or perspective that even the originator may not be completely aware of.

Although it may require considerable effort for a scientist to refine and describe his new idea in a language shared by other scientists, the impact of the new orientation or idea on the written theory is crucial, since the written theory is only a reflection of the new orientation or idea. The purpose of this chapter is to describe several types of "new ideas" and try to demonstrate what an orientation or conceptualization entails, even if only intuitively. Basically, the newness of a new idea can only be appreciated if one is aware of the scope and quality of the old ideas that prevailed before the new idea was introduced. It is convenient to classify "new ideas," according to their degree of "newness," into three types: Kuhn paradigms, paradigms, and paradigm variations. Each will be discussed separately.

KUHN PARADIGMS

The most dramatic type of new idea will be referred to as a "Kuhn paradigm," since it was originally suggested by T. S. Kuhn (1962). It has the following features:

(1) It represents a radically new conceptualization of the phenomena;
(2) It suggests a new research strategy or methodological procedure for gathering empirical evidence to support the paradigm;
(3) It tends to suggest new problems for solution;
(4) Application of the new paradigm frequently explains phenomena that previous paradigms were unable to explain.

In sum, a Kuhn paradigm includes not only a unique and unprecedented orientation toward the phenomena, a dramatic break with past and existing orientations, but also involves a major shift in research strategy (often including new research techniques) as well. Kuhn refers to his paradigms as "scientific revolutions."

There are a number of examples of Kuhn paradigms from the physical and biological sciences described by Kuhn (1962). Two will be presented here. Prior to Newton, the fall of objects to the earth was explained by objects seeking their "natural" resting place, the surface of the earth. When an apple was separated from a tree, it would seek its "natural" resting place and move toward or "fall" to the ground. Newton suggested a new way of describing the same phenomenon, as the attraction of two objects for each other. When the apple is separated from the tree, the earth is attracted to and moves toward the apple as well as the apple being attracted to the earth. Because of the difference in mass, the apple travels farther than the earth. Although this may appear at first to be a small difference in perspective, the new theory involves a dramatically different orientation when applied to the movement of planets in the solar system (where the mass of the bodies is more equal), and it was several hundred years before the new orientation was completely adopted.

Before Darwin, one of the dominant explanations of the apparent "harmony" between living forms and their environment, referring to the way living forms seemed to be so well adapted to their physical situations, was that each living thing was part of a master plan (conceived and executed by the Almighty) designed to produce "ideal" specimens of each species. Darwin suggested that this "harmonious relationship" could be explained by two earthly processes: the slight variations in living forms, continuously produced by natural genetic processes during reproduction, and the tendency of those living forms more suited to the environment to survive or to reproduce more frequently than those less suited to the environment. Clearly, Darwin's conceptualization was a dramatic break with past views, a scientific revolution that even today has not been universally accepted.

Perhaps because there has been less agreement among social scientists as to the nature of existing paradigms, or views of the phenomena, it is slightly more difficult to select well-known examples of Kuhn paradigms in the social sciences. Marx's conception of society, Weber's or Durkheim's strategy of analysis, the Cooley-Mead orientation about the development and maintenance of a self-concept, or Keynes's model of an industrial economic system could be called Kuhn paradigms. However, it is clear that one conceptualization, Freud's theory of personality, has all the characteristics of a Kuhn paradigm. Here is a brief summary of Freud's ideas, followed by a discussion of how they meet the criteria of a Kuhn paradigm.

EXAMPLE: *Freud's Theory of Personality*[1]

Freud assumed that *all* of an individual's sense impressions, both real and imaginary (fantasy), and his behavior were *caused.* His goal was to develop (Freud felt he was discovering) a single, integrated conceptualization of an individual's personality that would explain all of his behavior and sense impressions.

Instincts and Energy

Freud considered that all individuals have a number of physiological instincts or needs that must be satisfied. However, once satisfied, they continue to recur, placing the individual in a continuous cycle, as follows:

Need Arousal → Tension Increase → Satisfaction of Need → Tension Decrease → Need Arousal → Etc.

1. This summary is based on three other summaries, Blum (1966), Hall and Lindzey (1957), and Rapaport (1951).

Along with these needs, the physiological system also provides the individual with the energy necessary to satisfy these needs, including the energy required for the operation of mental processes —psychic energy.

All needs can be "satisfied" in two ways, either by achieving the required goal in reality (e.g., a hungry man eats food) or by creating images of the required goal (e.g., a hungry man dreams of food). Most of the basic physiological needs such as the need for breath, for drink, for food, for urination, and for defecation must be readily accommodated, and individuals tend to satisfy these needs in much the same ways. Therefore, there is little variation among individuals in the way in which these needs affect them or their "personality."

However, two other needs, sex and aggression (reflecting the "death wish"), may be satisfied in a number of ways, and individuals may vary considerably in the way in which they "satisfy" these needs. Personality development is considered to be a process whereby the individual learns how to satisfy these two needs, generally in ways that are acceptable to his society.

Structure of the Personality

Freud classified mental processes in two ways. The first classification is the division between conscious and unconscious thoughts or processes. The largest part, more than 80–90 percent, of mental activity is considered to be unconscious and cannot be observed by ordinary techniques.

The second classification is into the three systems of the id, ego, and superego. The *id,* entirely unconscious, is considered the source of all needs and psychic energy. The other systems, ego and superego, are required to satisfy the needs or drives of the id with real or imaginary sense impressions. The *superego,* largely a conscious system, represents the traditional values and ideals of society, considered to be transmitted to the individual during socialization, primarily by his parents. Its major purpose is to guide behavior designed to satisfy the id's needs in accordance with societal standards. The *ego,* which evolves from the id, is an administrator, attempting to satisfy the needs of the id with real or imaginary sense impressions in a manner acceptable to the superego. The ego is considered to have the capacity to differentiate between fantasy and reality, a distinction the id cannot make.

Personality Development

Personality development is considered a learning experience whereby the individual discovers how to satisfy the needs of the id, either through behavior acceptable to the superego (and, hence, society) or through fantasy. This is considered to occur primarily before the individual reaches the age of five in three stages, the oral, anal, and phallic, as sexually related interests change with maturity. Improper or inappropriate training at any of these stages will result in an "unhealthy" adult, with an oral, anal, or phallic type of personality, whereas appropriate training will result in a "healthy" adult, with a genital type of personality.

Research Procedure

Because Freud considered most mental activity to be unconscious, he used research procedures that enabled him to observe unconscious processes. Two of the most widely used techniques were free association, in which the individual talks about any number of topics that enter his mind in whatever order they occur to him, and the analysis of dreams, in which the individual describes his dreams and fantasies. Freud would collect large volumes of such material from his patients and then reconstruct this material so that it was both *internally* consistent and, of course, consistent with his conception of the personality structure. His goal was to provide an integrated and logical organization of the patient's mental activity that incorporated *all* of his dreams, fantasies, conscious cognitions, and, ultimately, his behavior. This thoroughness is a far cry from the glib interpretations based on isolated incidents one often hears from amateur psychoanalysts or mass-media interpretations of psychoanalysis.

There seems to be little question that Freud's view of individual personality and the research techniques associated with psychoanalysis qualify the theory as a Kuhn paradigm. Freud's reconceptualization of personality, particularly the importance of unconscious mental activity and its development, involving the concept of infant sexuality, represented a dramatic break with the past. The new research and therapy techniques of psychoanalysis were crucial to the development, testing, and application of the theory. His theory was a noble attempt to explain the importance of dreams and fantasies, imaginary sense impressions, that were not previously considered of importance. Finally, a large number of unsolved problems have been generated by the psychodynamic or Freudian theories, particularly with respect to the relationship of infant socialization to the adult personality.

PARADIGMS

If a new orientation is a less dramatic break with the past than, for instance, were those of Newton, Darwin, or Freud, it can be called a "paradigm" with the following characteristics:

(1) The conceptualization represents a unique description of the phenomena, but a *dramatic* new orientation or "world view" is absent;
(2) Although new research strategies may be suggested, dramatic new procedures or methodologies are absent;
(3) The new conceptualization may suggest new research questions;
(4) The new conceptualization may explain events previously unexplained.

Basically, a paradigm and a Kuhn paradigm differ in degree; the Kuhn paradigm represents a dramatic change from past orientations, whereas a paradigm represents a definite shift in orientation but less than a "scientific revolution."

Most "theories" or "orientations" in social science can be considered to be paradigms. The following list of examples could be extended:

Theories of cognitive balance
Theories of social exchange
Utility theory
Pluralistic conceptions of status structure
Elitist conceptions of status structure
Weber's conception of bureaucracies as goal-oriented mechanical systems
Barnard's conception of organizations as survival-oriented social systems

Stimulus–response learning theory
Information-processing models of cognitive processes
Structural–functional strategies.

Two examples will be described in detail.

EXAMPLE: *Heider's Balance Theory*

The first person to suggest that it might be useful to consider an individual's cognitive organization in terms of a tendency to prefer balanced cognitive systems was Heider (1946; 1958). Rather than attempting to understand and explain the entire mental structure of an individual, as in Freud's formulation, Heider suggested that one small part be selected for analysis, focusing on an individual's attitudes (positive or negative) toward other individuals or objects and his perception of the relationship between these other cognitions (i.e., the other individuals and objects).

For example, consider the cognitive structure of one individual, Peter. Peter's cognitive structure consists of three objects: himself, another person, Oscar, and an object, a xylophone. Three relationships among these cognitions complete the cognitive structure: Peter's attitude toward Oscar, Peter's attitude toward the xylophone, and Peter's perception of Oscar's attitude toward the xylophone. If each of these relationships is considered to be either positive or negative, then Peter could have any one of eight types of cognitive structures (see Fig. 1).

Heider's procedure was to examine each of these eight types of cognitive structures and speculate on his reaction to each, comfortable or uncomfortable, assuming he was in Peter's position. He felt more comfortable with the structures he labeled "balanced" than with structures labeled "imbalanced." Heider then suggested that if an individual had an imbalanced cognitive structure he would experience psychological tension, an uncomfortable state. The individual should then take action to avoid this uncomfortable tension, perhaps by changing some aspect of his cognitive structure, such as a change in attitude or perception. However, in most cases several changes could be made to balance a cognitive structure, and Heider only suggested that changes would occur, not which of several alternatives would actually develop to reduce imbalance (see Berger *et al.,* 1962, for a thorough analysis of Heider's original article).

FIGURE 1. ALL OF PETER'S POSSIBLE COGNITIVE STRUCTURES

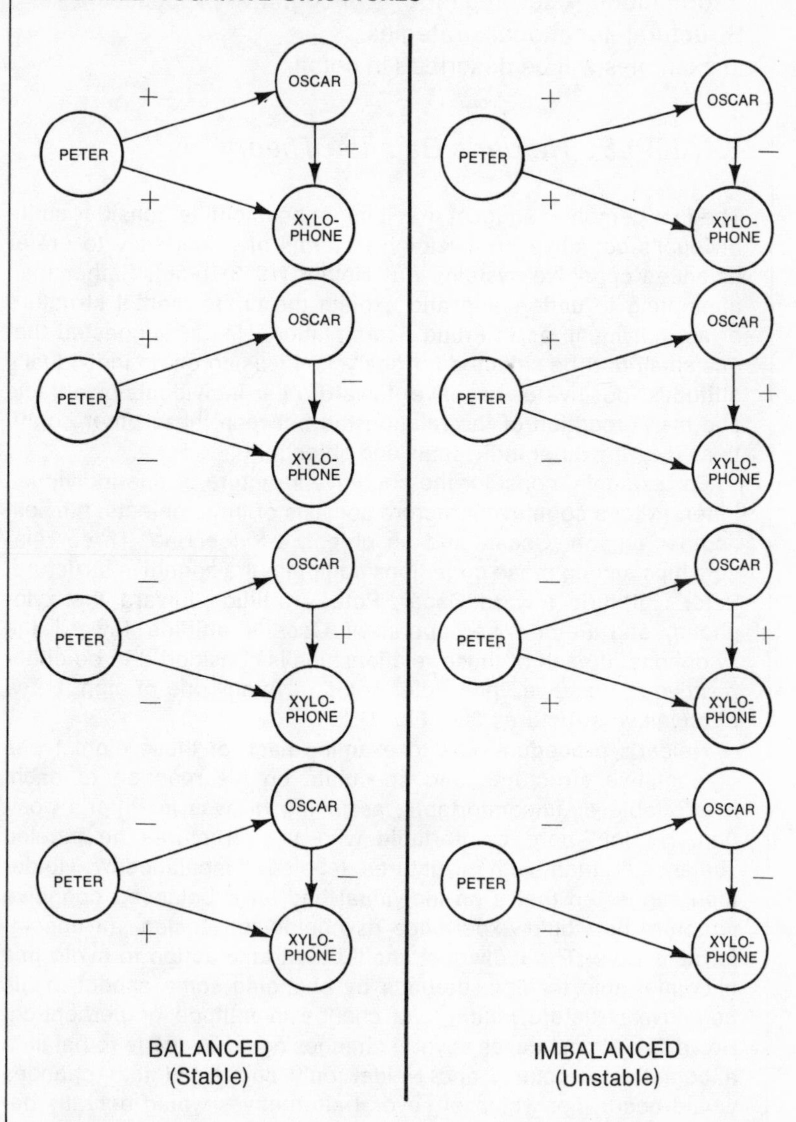

Note: Circles represent cognitions; arrows represent attitudes or orientations, perceived or experienced by Peter; + represents a positive attitude and − a negative attitude.

Heider's orientation meets all the criteria of a paradigm. His suggestion that it is fruitful to analyze subparts of an individual's cognitive structure clearly suggests a unique view of the phenomenon. As is often the case, his paradigm actually defines a new phenomenon, cognitive structures. However, considerable work on cognitive organization preceded Heider's writing, particularly by others trained in Germany (Lewin, 1936) in the Gestaltist tradition (Kohler, 1929). Thus, Heider's orientation did not represent the dramatic break with past ideas associated with "scientific revolutions" or Kuhn paradigms.

However, Heider's paradigm suggests new research strategies designed to identify cognitive structures and their degree of balance or imbalance. New research questions are generated by his conceptualization: How does an imbalanced cognitive structure affect the individual's behavior? How is an imbalanced cognitive structure changed so that balance is achieved? It is not immediately clear that Heider's paradigm was able to explain phenomena that were previously unexplained, but not all paradigms, even Kuhn paradigms, represent dramatic solutions to existing problems. The basic elements of a paradigm are contained in Heider's ideas.

EXAMPLE: *Two Conceptions of Status Structures: Elitist and Pluralistic*

It is generally conceded that social systems are composed of individuals that differ in a number of characteristics, such as education, occupation, income, ethnicity, etc. In attempting to describe social systems, it has been considered useful to suggest that the individual members differ in terms of their status, or prestige and influence over others, within the social system—the status structure. Two alternative conceptions of the nature of the status structure within a social system are currently employed to explain and describe status- (and power-) related phenomena. Both concep- tions involve the measurement of differences in individual cha acteristics, such as different types of occupations, in order measure individual differences in status and power; but they in how these characteristics are used to identify the status ture, a characteristic of the social system.

Elitist Conception: Status Structure as a Set of Social Classes

In this conceptualization, all of the factors that define an individual's status are considered to be positively correlated, i.e., all highly valued characteristics occur together, and the social system is seen as a set of groups or classes that can be arranged in a hierarchy (see Fig. 2). Those in the "higher" social classes are

FIGURE 2. TWO CONCEPTIONS OF STATUS STRUCTURE

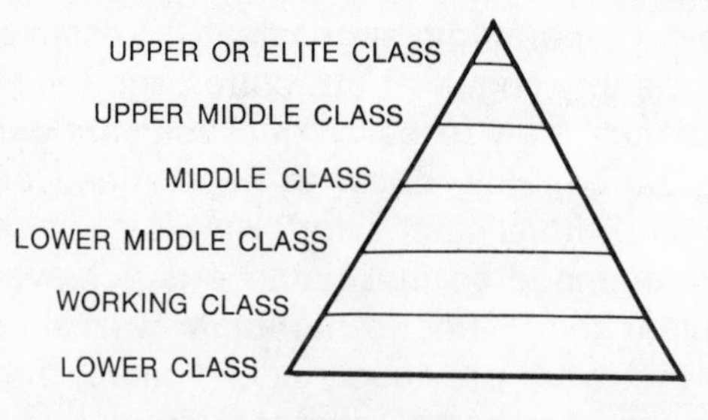

SOCIAL CLASS OR ELITIST CONCEPTION

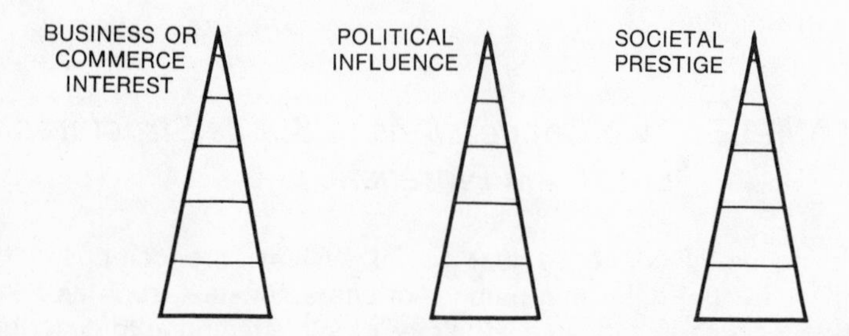

INDEPENDENT HIERARCHIES OR PLURALISTIC CONCEPTION

considered to have better jobs and more influence in the world of commerce, more education, higher incomes, more political influence, higher ethnic ranks, etc. Those in the "lower" classes are considered to have poor jobs, little influence in the world of commerce, little education, low incomes, little political influence, "lower" ethnic rank, etc. Individuals in the "middle" classes are considered to be intermediate in all these characteristics.

When working with this conception, social scientists are led to ask questions about the tendency of individuals to move from class to class, both up and down, and the location of the boundaries, as well as the relative size of each social class. If research indicates that some individuals have inconsistent sets of characteristics, i.e., advanced education and low income or little education and much political influence, then this is considered a temporary aberration that will eventually disappear as the individual settles into his "appropriate" class or as the social system stabilizes. Because some have suggested that the "upper class" is composed of an elite few that have political control of the entire social system (be it a city, state, or nation), this is sometimes referred to as the "elitist" conception of a status structure. One of the most popular research activities of those using this conception of a status structure is to identify the "elite group" or "upper class" in a social system who are assumed to monopolize power and influence in the system.

Pluralistic Structure: Status Structure as a Set of Independent Hierarchies

In this conceptualization, each of the different dimensions is seen as an independent prestige and influence rank order, and the individual may have completely different ranks in two or more characteristics. A particular individual may be high in a "political influence" rank order and relatively low in "influence in the business community." Because this conceptualization of a status structure suggests a number or "plurality" of influence dimensions, or power centers, it is sometimes called a *pluralistic* conception of status structures.

When using the pluralistic conceptualization of a status structure, social scientists tend to focus on how an individual transfers high rank in one hierarchy, such as income or education, into high rank in another hierarchy, such as political influence. Attention is also given to the effect on an individual that occupies different ranks in several hierarchies, such as high education and low income, the status inconsistency problem. Perhaps the research problem most directly related to the pluralistic phenomena is the attempt to identify the individuals that make the decisions, wield the power, in different areas, such as in commerce, education, health, public works, etc.

It seems clear that these two conceptions of status structures represent different orientations, explain different types of phenomena, suggest different research questions, and suggest different types of research methodology. (The elitist conception suggests that it is fruitful to attempt to identify each social class; the top or elite class has been popular. The pluralistic conception suggests that individuals can be identified with respect to their status or influence on different issues—political, business, educational, etc.) There is reason to think that, in this particular case, the pluralistic paradigm has gained acceptance because of data collected that were inconsistent with the elitist paradigm—a case in which a new paradigm gains prominence because of the inadequacies of the existing paradigms. However, both conceptualizations are presently considered useful for the study of status structures, in much the same way that two conceptions of light, as waves and particles, presently coexist in physics.

Each of the examples discussed or suggested as paradigms provides a different way of conceptualizing or describing certain phenomena. They are all important in attempting to understand and explain the subject matter of social science, which is in part defined by the "world view" in the paradigms. Paradigms differ from Kuhn paradigms only in degree; when introduced they provided an orientation that is less than a "scientific revolution." Most of the activity in social science is related to issues or questions "generated" by paradigms; each is embraced by a subgroup of social scientists, at least for the duration of the research project.

PARADIGM VARIATIONS

Once a conceptualization or orientation on the level of a paradigm or a Kuhn paradigm has been proposed, there are often a large number of details or refinements that are ambiguous or unspecified. Frequently there are several alternatives available in specifying the details of the paradigm, each resulting in slightly different variations in the original conceptualization. These slight variations in paradigms, Kuhn or otherwise, are called "paradigm variations" and are considered to offer refinement of details or variations in emphasis, not changes in the basic conceptualization of phenomena associated with the paradigm.

The examples of paradigm variations in social science are almost endless, since every book or article seems to suggest a new variation of an existing paradigm. However, some paradigm variations are better known than others, and two examples will be presented here: (1) variations of psychodynamic personality theory based on Freud's original conceptualization, and (2) variations of cognitive balance or cognitive consistency models extending Heider's original conceptualization.

EXAMPLE: *Variations on the Freudian Conception of Personality*

A large number of variations on Freud's original conceptualization have appeared; some of the more prominent will be briefly described here (taken from Blum, 1966, pp. 13–21). In all of these paradigm variations considerable importance is given to unconscious mental activity as an influence on behavior and sense impressions.

Alfred Adler suggests that much of adult behavior is better explained by a reaction against feelings of inferiority. These feelings develop naturally in any child because of his size and helplessness in relation to his parents. All adults attempt to compensate for this inferiority complex by attempting to dominate or influence those around them.

Carl Jung suggests that unconscious processes are divided into the personal (reflecting the individual's personal experiences) and the collective (reflecting "cultural" concepts and themes). He also suggests that individuals may be fruitfully classified in terms of which basic psychological function they emphasize (thinking, feeling, sensing, or intuiting) or basic attitude toward the world (extroverted or introverted).

Otto Rank suggests that the life goal of individuals is to achieve individuality as well as to regain the feeling of contentment initially experienced in the womb. All separations, reminiscent of the birth trauma and separation from the womb, are viewed as threatening. All individuals are considered to have a positive, building aspect of the self, the *will,* which controls and utilizes "impulses" creatively. Depending on early socialization by the parents, the "will impulses" may create an average, neurotic, or creative adult.

The following group of *neo-Freudians* place stronger emphasis on cultural influences on personality.

Karen Horney suggests that children can cope with their early environment by moving toward, against, or away from people. Depending on which strategy the child takes, he will become an adult with a compliant, aggressive, or detached type of personality.

Erich Fromm suggests that the more an individual loves himself, the greater will be his capacity for loving others, in direct contrast to Freud's view that one could intensify love for self or others, but not both simultaneously. Fromm also prefers to divide the conscious into the "authoritarian conscious," the internalized voice of external authority, and the "humanistic conscious," the individual's real feelings of self-interest and integrity. Depending on the home atmosphere prevalent during childhood socialization, an adult's character type will be receptive, exploitative, hoarding, marketing, or productive.

Harry Stack Sullivan suggests that a "power motive," an attempt to control others, operates from birth to compensate for an inner sense of helplessness. As the individual grows up, he passes through the prototaxic and parataxic modes of experience until he achieves the syntaxic mode of experience, which is free from distortion.

There is another group, the *ego-psychologists,* including *Heinz Hartmann, Ernst Kris, Rudolph Loewenstein,* and *Erik Erikson,* who have shifted emphasis toward increasing the functions of the ego, as an autonomous or independent set of mental processes, and away from the "deeper" (more unconscious) processes of the id. They even suggest that the id and ego develop simultaneously, rather than the ego from the id, as Freud suggested.

These variations (1) consider there is a cause for all behavior, (2) attempt to explain both behavior and all sense impressions, real and imaginary, with the same conceptualization, (3) generally consider the unconscious mental processes and unconscious needs important, and (4) consider early childhood experiences important in the development of adult personality —all part of Freud's original conceptualization. Because they share these important features with Freud's original paradigm, they can be considered paradigm variations.

EXAMPLE: *Variations on Heider's Balance Theory*

Four variations on Heider's paradigm will be described. Others may be found in Secord and Backman (1964, p. 110).

Osgood and Tannenbaum (1955)

These men chose to focus on one special type of situation, among the many suggested by Heider. They selected a cognitive structure with two individuals and an issue. One individual is the person whose cognitive structure is being studied, again referred to as Peter. The other is a source of opinions or attitudes that is well known to Peter and can be a newspaper, politician, well-known "intellectual," or some similar entity, and is referred to as the Source. Any topic toward which Peter may hold a positive or negative attitude is referred to as the Issue. In contrast to Heider's classification of attitudes as positive or negative, favorable or unfavorable, Osgood and Tannenbaum classify attitudes into seven degrees, from −3 to +3. This more precise measurement of attitudes is what they gain from focusing on only one of many situations.

Osgood and Tannenbaum are interested in a situation in which they can measure the changes in attitudes that will occur if an incongruent, or imbalanced, cognitive structure occurs. They actually produced imbalanced, or incongruent, cognitive structures in an experimental situation. The procedure, diagrammed in Figure 3, is as follows:

Step I Select an Issue on which Peter has a neutral attitude. Select a Source (of opinions) toward which Peter is positively or negatively oriented.

Step II Inform Peter that the Source has either a positive or negative opinion with respect to the Issue, in such a way as to create an incongruent (imbalanced) cognitive structure—if Peter believes the experimenter.

Step III Assuming that Peter will prefer a congruent or balanced cognitive structure, measurement of Peter's attitudes toward the Issue and the Source should indicate changes in attitudes.

FIGURE 3. FOUR EXAMPLES OF A CHANGE IN ATTITUDE TO ACHIEVE A MORE BALANCED COGNITIVE STRUCTURE (after Osgood and Tannenbaum, 1955)

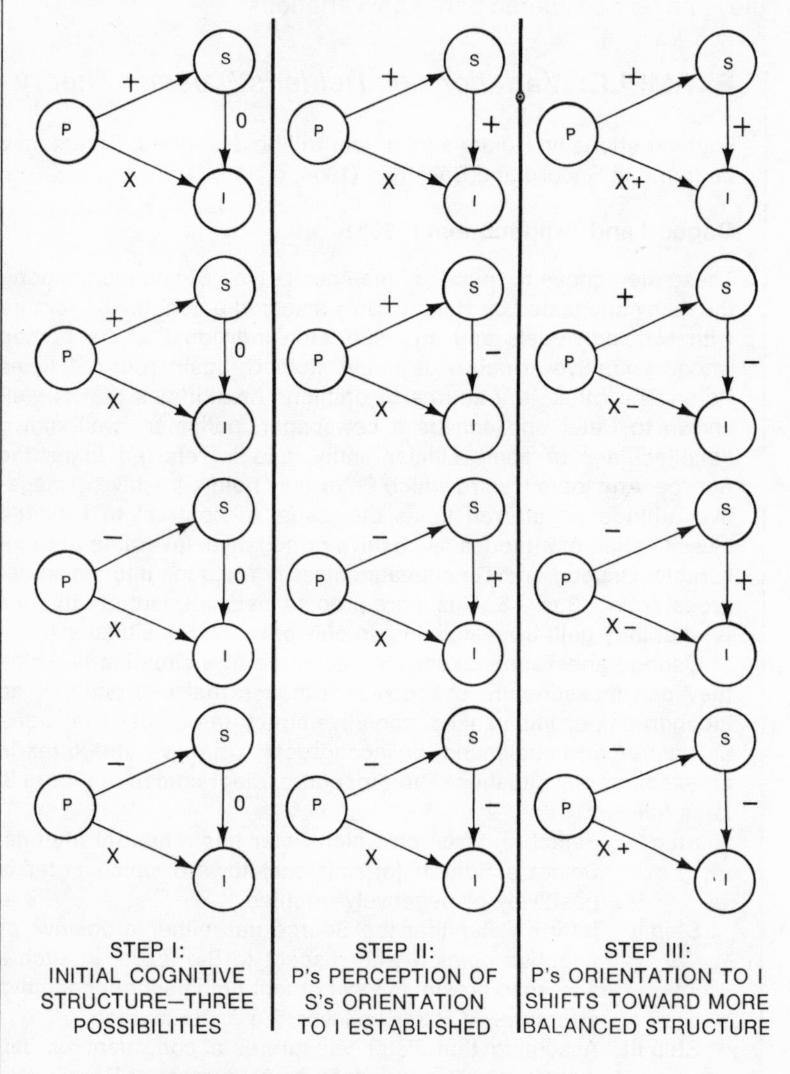

P = Peter, S = Source, I = Issue
+ = positive, — = negative, 0 = unknown, X+ = positive change,
X— = negative change

The congruent cognitive structures in Step III of Figure 3 can be compared with the balanced cognitive structures in Figure 1 (p. 28).

This represents only one set of situations that can be analyzed with the Osgood and Tannenbaum formulation in this special three-entity case. Since they work with the degree of attitudes, a large number of other situations are possible, each dealing with the change of degree of attitude (e.g., from −1 to −2) rather than from positive to negative.

Rosenberg (1956)

Rosenberg is concerned with cognitive structures from the individual's point of view, in a manner similar to Heider, but he focuses on different types of cognition. Heider considered them all of equal importance; Rosenberg classifies them into two groups. The first is the group of *values,* cognitions (1) on which the individual has a very strong orientation, or affect, either positive or negative, and (2) which the individual considers an important goal in life or a desirable state of affairs (i.e., people sticking to their own groups, having a steady income, being well educated, etc.). The second group is the group of *objects,* cognitions toward which the person may have a positive or negative orientation, but which are not nearly as important as values (i.e., allowing members of the Communist Party to speak in public, owning a Ford, etc.).

Rosenberg next considers how an individual might perceive the relationship between an object and a value in his cognitive structure. He suggests that individuals may perceive two types of relationships between objects and values. One relationship, which he calls "instrumentality," specifies the degree to which the object is seen as instrumental, or useful, for helping to achieve the value. In other words, would a positive orientation toward the object help the achievement of the favored value? Alternatively, would a negative orientation toward the object help prevent the achievement of the unfavored value? The second relationship is "relevance," the degree to which the object is seen as relevant to the attainment of important values, both favored and unfavored. In other words, is the object in any way related to the value?

While Rosenberg has adopted the notion that the cognitive structure will be balanced, he considers new ways of describing the relationship between cognitions in the structure. For instance, if a person feels that one of his most important values is "people being strongly patriotic," then it is likely that he will perceive "allowing members of the Communist Party to address the public" as both

(1) very relevant to "being patriotic," and (2) detrimental to the achievement of "being patriotic." Therefore, the individual would be expected to have a negative attitude toward "allowing members of the Communist Party to address the public." On the other hand, if a person (even the same person) had a positive orientation toward the value of "having interesting work to do," it is less likely he would see the object, "allowing members of the Communist Party to address the public," as either relevant or instrumental to "having interesting work to do." Therefore, there is no reason to expect the orientation toward the value, "interesting work," to be systematically related to the attitude toward the object, "allowing members of the Communist Party to address the public."

Rosenberg's variation maintains several important features of Heider's original conceptualization. He looks at the cognitive structure from the individual's perspective, he stresses positive and negative orientations toward cognitions, and he assumes that only certain types of cognitive structures, the balanced ones, will be stable and identifiable by the researcher.

Cognitive Dissonance (Festinger, 1957)

This version of Heider's idea also uses a slightly different notion of a cognition and the relationship between cognitions. *Cognitions* are assumed to be more abstract than simply a person, an object, or a value, and are considered to have a number of characteristics. For instance, the cognition "trip to Europe" may have the positive characteristics of an exciting way to spend two months and an experience that may be shared with one's friends afterwards. It may have the negative characteristics of expense, preventing the alternative use of the money (such as purchasing a luxury car), or removing one from comfortable social relations for several months. All of these characteristics, positive and negative, are considered part of the cognition "trip to Europe."

Two, or more, cognitions may be considered in terms of their "psychological logical" relationship to one another. This "psychological logical" relationship is a set of ideas, presumably culturally defined, about what kinds of cognitions should "fit" together—be associated with one another. Although the actual "psychological logic" is not clearly specified in the cognitive dissonance literature, it can be considered as having three states: (1) consonance—two or more cognitions that "fit" together, according to psychological logic; (2) dissonance—two or more cognitions that do not "fit" together, according to psychological logic; and (3) irrelevance—two or more cognitions that share no relationship with one another, according to psychological logic.

The basic assumption is that two or more cognitions should "fit" together, be consonant, and if they are not consonant, and not irrelevant, they will be dissonant. Dissonant cognitions are considered to be uncomfortable and to cause psychological tension. In order to reduce this tension, a person may take a number of actions in order to change his cognitive structure and to reduce or eliminate the dissonance between cognitions. One way for an individual to change his cognitive structure is to change his perception of certain events in reality.

For instance, if a person is considering the decision to make a major purchase, such as a car, before the purchase he may consider car *X* and car *Y* as quite different, but their overall attractiveness may be about equal. After he has purchased one of the brands, he must achieve consonance between several important cognitions: "I am a rational purchaser and buy products on their merits," "I own brand *X*," and "brand *X* and brand *Y* are equally attractive." If these three cognitions are considered to be "illogical" according to "psychological logic," then the individual is in a state of cognitive dissonance and will experience uncomfortable psychological tension. In order to reduce the tension the individual may reduce the dissonance among the cognitions by changing the cognition "brand *X* and brand *Y* are equally attractive" to "brand *X* is much more attractive than brand *Y*." This could be done by emphasizing as desirable the advantages of brand *X*, such as economy and ease of maintenance, and deemphasizing as undesirable the advantages of brand *Y*, such as acceleration, speed, or style. The individual may even take actions designed to help him "convince himself" that "brand *X* is more attractive than brand *Y*" by seeking advertisements for brand *X* and avoiding advertisements for brand *Y* (see Ehrlich *et al.*, 1957).

Newcomb (1953)

Newcomb focuses on a situation that is slightly different from Heider's paradigm and the preceding variations. He focuses on a system of two actors, *A* and *B*, and an object, *X*, but includes the social relationships between the two actors in his analysis. Instead of concentrating on the organization of *one* person's cognitive

FIGURE 4. THE A–B–X SOCIAL SYSTEM AND THE COGNITIONS OF THE MEMBERS: FOUR EXAMPLES (after Newcomb, 1953)

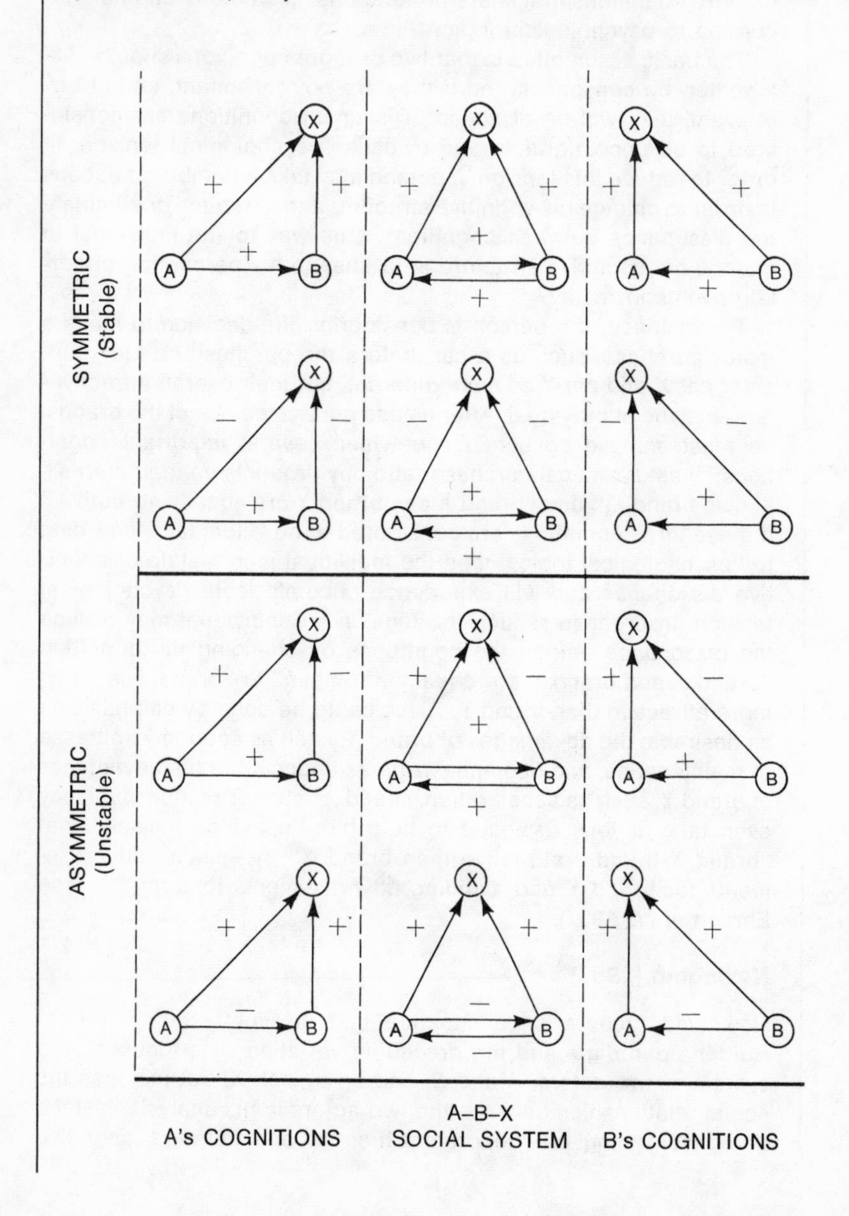

structure, he also treats the actual relationships between the two actors.

The basic social system is shown in the middle column in Figure 4 and shows four relationships, *A*'s attitude toward *X*, *B*'s attitude toward *X*, *A*'s attitude toward *B*, and *B*'s attitude toward *A*. Notice that the symmetric social systems, if accurately perceived by *A* and *B*, lead to balanced cognitive structures (compare them with the balanced structures in Fig. 1, p. 28). Two examples of asymmetric social systems are presented. Notice that they lead to imbalanced cognitive structures on the part of the actors, *A* and *B*, if the actors accurately perceive the relationships.

Newcomb's variation is unique in that it focuses on the behavior between the actors, *A* and *B*, in an asymmetric social system rather than just changes in cognitions. Newcomb suggests that if there is an asymmetric system, the actors may discuss their attitudes toward the object, *X*, in order to come to an agreement about their attitude toward *X*. For example, if one person finds that they like another very much (*A* likes *B*), perceives that the other likes him (*A* perceives that *B* likes *A*), likes sport cars (*A* likes *X*), and perceives that the other person does *not* like sport cars (*A* perceives that *B* dislikes *X*), Newcomb's formulation would suggest that the person would "communicate" with the other. *A* would talk to *B* in an attempt to change *B*'s opinion about sport cars. If the other person perceives the situation in the same fashion, he may "communicate" to *A* an attempt to change his opinion about the sport cars, *X*. A lively discussion should ensue. It is for this reason that Newcomb called his formulation "A Theory of Communicative Acts."

Four variations of Heider's original paradigm have been discussed. They all have in common a complete or partial focus on the cognitive structure of a single individual as well as an assumption that "balanced" cognitive structures are more desirable or cause less "tension" than imbalanced, inconsistent, incongruent, dissonant, or asymmetric cognitive structures and that this will *cause* the individual to take action to reduce the imbalance, inconsistencies, incongruencies, dissonance, or asymmetry in such a way as to reduce the tension. Although the theories emphasize different reactions to the imbalanced cognitive structure—changes in attitude, changes in cognitive organization, or changes in behavior—the basic conceptualization in Heider's paradigm is reflected in all of these paradigm variations.

IDENTIFYING PARADIGMS

Although it is possible to determine which conceptualizations or orientations *have* become major paradigms, Kuhn or otherwise, it is often very difficult to determine which of the multitude of new ideas *will* become a new major paradigm. Paradigm variations are usually easy to identify because of rather direct and obvious links with existing paradigms, reflected in footnotes, bibliographies, and similarity of orientations.

Two factors seem to slow down the acceptance of new paradigms. First, existing scientific terminology is usually related to existing conceptualizations of phenomena; the more dramatic the change in conceptualization, the more difficult it would be to describe a new "concept" that is not part of the existing vocabulary. Second, the processes involved when a scientific community adopts any paradigm are both subtle and lengthy; some Kuhn paradigms (the major scientific revolutions) were not fully adopted for hundreds of years. It is unfortunate, but some individuals who create new paradigms never live to see the full adoption of their ideas by the scientific community.

In fact, there seems to be considerable evidence to suggest that when a dramatically new Kuhn paradigm, or major scientific revolution, is first introduced, it is often met with a great deal of skepticism and hostility (Barber, 1961). However, this should not be interpreted to mean that all ideas that are greeted with skepticism and hostility will eventually create a scientific revolution. This is more of an indication of the difficulty the scientific community has in separating the few new ideas that will prove to be useful from the large number of new ideas that offer little improvement over existing paradigms.

In other words, although new and radical conceptualizations of "real world" phenomena are constantly being proposed, only a small minority of these will eventually create scientific revolutions or even new paradigms useful for the goals of science.

CONCLUSION

A formal description, or "theory," only attempts to describe an idea, and the idea is the most important feature of any theory, formal or otherwise. But it should be emphasized that although the formal descriptions only reflect a "conceptualization," unless the "conceptualization" is explicitly described other scientists cannot understand it and probably will not adopt it. Unfortunately, it takes a great deal of work to describe any idea in a scientifically useful form, whether the idea is eventually adopted as useful (or good) or rejected as not useful (or bad).

This chapter provided a description of different types of new ideas, classified in terms of their "newness" and impact on science, and tried to present an intuitive feel for the concept of a "conceptualization," "orientation," or "paradigm." The remainder of the book will focus on the different parts of the formal description, suggest the steps one should take in constructing such a description, and discuss how scientists evaluate theories and decide which they will adopt.

3. Concepts

This chapter will be devoted to a discussion of four issues related to concepts: (1) their definition; (2) the difference between abstract and concrete concepts; (3) the relationship between abstract concepts, used in theoretical statements, and operational definitions, the procedures prescribed for the measurement of abstract concepts in concrete settings (the location of all empirical research); and (4) quantification of theoretical concepts and operational definitions.

If it is assumed that the goals of scientific knowledge are to provide a system of classification (a typology), explanations, predictions, and a sense of understanding, then it is clear that only the first of these goals, providing a typology, can be expressed by concepts alone. The remaining goals (explanations, predictions, and a sense of understanding) are expressed by statements—statements that contain scientific concepts. Therefore, for most of the purposes of science, concepts cannot be judged apart from their use in statements. In other words, the scientific value of concepts can only be judged in terms of the scientific utility of the statements containing them. However, concepts can be evaluated in terms of the clarity with which they are described—clarity measured by the degree of agreement among the users of the concept on its meaning.

DEFINITION OF CONCEPTS[1]

If an individual wishes to share his ideas with another, he must somehow convey his ideas. The only acceptable means for sending and receiving scientific messages is through the use of a written language—either natural or artificial, such as mathe-

1. See Hempel (1952) for a more complete discussion of definitions in empirical science.

matics. (Any other language could not be accurately transmitted from one generation of scientists to the next.) Therefore, the problem is one of insuring that the sender and receiver agree on the meaning of the symbols used to represent concepts.

There are two types of symbols or terms used in any language, natural or artificial. First, there are *primitive* symbols, those on which there is shared agreement as to their meaning but which cannot be described using other symbols or terms. Second, there are *derived* (*nominal*) symbols or terms, those that can be described by the use of primitive terms.

Consider the following examples:

Primitive terms (shared agreement exists among users):

X	Individual	Goal-oriented
Y	Interact	Social system
+ (operation performed with numerals)	Two or more	Formalized
	Regularly	Rules and procedures

Derived (*nominal*) *terms* (defined with primitive terms):

$Z = X + Y$	Group—two or more individuals that interact regularly	Formal organization—a goal-oriented social system with formalized rules and procedures

Either a derived term (*Z*, "group," "formal organization") or its definition, composed of primitive terms, expresses the same concept. The main advantage of the derived term is that its use is more efficient, requires less effort, than the set of words that compose the definition of the derived term. In other words, knowing the definition of a derived term lets one know how to avoid the derived term; the definition can always be used instead.

One recurrent problem in social science is the tendency for members of the audience to add meaning to words that have been carefully defined by the originator, particularly if that word is used for other concepts. Often this added meaning, unintended by the writer, can dramatically change the meaning of a statement. One solution to this problem is to use abstract symbols, invented words, words that are seldom encountered, or even Latin or Greek phrases to label a concept. However, such terms are often criticized for being sterile, abstract, or hard to read. As long as an audience is careless about their interpretations of terms and simultaneously insists they be entertained, there seems to be no solution to this problem.

Primitive terms are more difficult to convey to another because they are those terms that cannot be defined by other symbols of the language. Ultimately, their meanings can be conveyed only by indicating examples, and nonexamples, of the concept to which each term refers. When this occurs, the "writer" is providing the "reader" with an opportunity to experience a certain collection of sensory impressions (sights, sounds, smells, tactile sensations, tastes, etc.) that cannot be described, but only can be labeled by the primitive term.

For instance, a particular color, sound, or relationship between individuals may not be easily described, but it might be possible to present a procedure that would allow another to experience the sensations labeled as that sound, that color, or that relationship. The term "butterflies in my stomach" does not convey much meaning, but to describe the queasy feeling one gets on the first big hump of a roller coaster or going down in a fast elevator as "butterflies in my stomach" may allow some agreement on the meaning of the term among those who have had this experience. Perhaps the best example is the concept of a "sexy girl." Although some characteristics can be put into words, or "measurements," at some point quantitative descriptions fail and one can only indicate girls that are and are not "sexy."

In sum, the only way to "define" a primitive term, to insure that all users are relating the term to the same concept, is to point out instances of the concept and instances of not-the-concept so that another can experience the sensory impressions that are defined as the concept. For instance, in Chapter 2, in order to demonstrate the concept "paradigm," different types of paradigms were presented.

Because primitive terms are often related directly to shared impressions, they are based on several individuals sharing the same experiences. Often several scientists working with the same phenomenon may have similar experiences, and therefore similar impressions, not shared by others (scientists or nonscientists). This select group of scientists may therefore agree on the meaning of certain primitive terms, whereas other scientists not working with the phenomenon, or nonscientists, may not feel comfortable with these terms. This is why intersubjectivity refers to agreement among the relevant scientists, not to all possible audiences. However, the potential for sharing the experiences, the impressions, and hence the primitive terms, should always be present. In other words, anyone with the proper training should be able to share these special experiences and, therefore, the meaning of the primitive terms.

Two other types of definitions should be mentioned, dictionary and real. "Dictionary definitions" are essentially attempts to describe the concepts indicated by the terms (or words) of a natural language. With a little exploration one often finds that dictionary definitions are circular; looking up one word leads to a second which leads to a third which leads back to the first word. This is because the dictionary cannot be more explicit than the primitive terms of the natural language and ultimately must assume that the reader understands certain primitive terms. The abundance of pictures in a dictionary testifies to the attempts to provide the reader with a sensory impression (in this case visual) that cannot be conveyed, or at least not conveyed efficiently, with words.

"Real definitions" are definitions that describe the real "essence" or real "characteristics" of an object or phenomenon. This type of definition assumes that objects or phenomena have some real property (or properties) that can be discovered and, hence, described. A more recent approach is to assume that observers attribute characteristics to objects and phenomena, not that there is any hidden "reality" to be discovered. Consequently, real definitions are seen less often.

In conclusion, the most important feature of any scientific term, used to indicate a concept, is the degree of agreement about its meaning, agreement about the nature of the concept. Derived definitions are composed of primitive terms that refer to concepts shared by the relevant scientists. Achieving agreement among the audience on the meaning of a term is more important than the actual form of the definition.

ABSTRACT VS. CONCRETE CONCEPTS

The term "abstract" refers to two different kinds of characteristics of concepts. The most frequent use is in the comparison of abstract and concrete concepts. In this use, abstract concepts are those concepts that are completely independent of a specific time or place. In other words, an abstract concept is not related to any unique spatial (location) or temporal (historical time) setting. If a concept is specific to a particular time or place, then it is considered concrete.[2]

The following are examples of abstract and concrete concepts.

ABSTRACT	CONCRETE	DIFFERENCE—CONCRETE IS:
Temperature	Temperature of the sun	Specific to a location
	The temperature of the earth on July 6, 1867	Specific to a location and a historical time
Three days	Dec. 4, 1967 to Dec. 6, 1967	Specific to a historical time
Attitude	What Harry thinks of the President	Specific to a particular individual
	What the President thinks of Harry	
Social system	The United States	Related to particular social system
	General Motors	
Face-to-face group	Joe, Harry, Pete, and Bob	Specific group of people
	Jones family	

In each case the meaning of the concrete concept is included within the meaning of the associated abstract concept; the concrete events are instances of abstract concepts.

A statement is basically the description of a relationship between two or more concepts. Since relations are always considered abstract, independent of any content, the "level of abstraction" of a statement will depend on the "level of abstraction" of the concepts. If the concepts are abstract, the state-

2. Some writers (Popper, 1957) shorten the phrase "applicable to a single instance" to the word "singular" in referring to concrete concepts or statements and shorten the phrase "universally applicable" to the word "universal" in referring to abstract concepts or statements.

ment will be abstract; if the concepts are concrete, the statement will be concrete.

Abstractness has several meanings; only one (independence of time and space) has been discussed. Two concepts may be independent of time and space and one statement may still be more abstract than the other. If one concept is included within the meaning of another, the second, or more general, concept is considered the more abstract. For instance, consider the concept "sentiment," an emotional disposition directed toward another, and the concept "liking," a positive feeling for another. The concept "sentiment" may be considered as including "liking" (along with "love," "respect," "admiration," and others). In this example, "sentiment" is the more abstract concept, for it encompasses the meaning of "liking."

Can a concept be too abstract? The answer is yes. For example, consider again the concept "sentiment." Love, respect, liking, and esteem may all be considered different types of sentiments. Although it is true that liking and love might generally be positively correlated (the same person is both liked and loved) or not correlated (a person liked is not loved), they will probably not be negatively correlated (a person is disliked and loved). On the other hand, a person might be disliked and respected, and the two attitudes would be negatively correlated. In such a situation, it is difficult to determine just what sentiment exists between the first person and the second, since disliking is negative and respect positive. The theoretical concept "sentiment" encompasses too much; it is too abstract, and this could lead to confusion in determining its existence.

This seems to lead rather directly to a criterion for determining whether a theoretical concept is too abstract or too broad. Unfortunately, this criterion rests on understanding the meaning of "operational definition," discussed below. In simplest terms, an operational definition is a set of instructions, independent of time and space (or, abstract), that describes what operations allow one to determine if a concept exists in a particular situation. Generally, theoretical concepts are considered as more abstract than operational definitions, and several operational definitions may be "indicators" of a single theoretical concept.

Theoretical Concept—Any concept more abstract than an operational definition, or procedure for measurement, that is considered to be part of a theory or potentially useful for inclusion in a theory.

This means that "sentiment," "liking," "respect," etc., would all qualify as theoretical concepts, since they are more abstract than a measurement procedure.

A theoretical concept can be considered too abstract if one set of instructions for identifying instances of the concept produces results that are inconsistent with the results of a second set of instructions for identifying the concept. They need not be positively correlated, but they should not be negatively correlated. If such negative correlations occur, then it will be difficult to agree on just when occurrences of the concept have been identified.

If inconsistency in identifying instances of the theoretical concept occurs, the difficulty can be attributed either to the operational definitions or to the theoretical concept. If the theoretical concept is considered at fault, it can be redefined so that one of the operational definitions is no longer considered an indicator of the existence of the event described by the concept; or it can be divided into two theoretical concepts, each related to one of the inconsistent operational definitions.

In summary, a theoretical concept should not be concrete,[3] that is, it should not be related to a particular spatial or temporal setting, but it should not be so abstract that there is confusion in identifying instances of the concept. Determining the most useful level of abstraction for theoretical concepts, above the concrete level, is largely a matter of judgment. As empirical evidence on the usefulness of a theoretical concept accumulates, it becomes easier to determine if the "right level" of abstraction has been reached.

3. Willer and Webster (1970) have suggested that social science has been retarded by the use of theoretical concepts at too concrete a level, preventing a cumulation of empirical evidence related to social phenomena. Results from different concrete situations are reported using different concrete terms, making it difficult to discover if they are similar at an abstract level. The authors suggest that research findings be reported at a more abstract level, similar to the level presented in the examples in this book, using a common set of theoretical concepts. Just how a common set of theoretical concepts, accepted by all social scientists as useful for describing all phenomena, is to be developed is not made clear.

CONCEPT MEASUREMENT

One of the important characteristics that a scientific statement should have is empirical relevance: it should be possible to compare the statement to some phenomenon or phenomena. This is generally done by identifying instances of the theoretical concepts in concrete settings. This means that it is important that some of the theoretical concepts be related to sensory impressions in concrete situations. However, it is not necessary that *all* of the concepts in a set of statements, or theory, be measurable. After all, *no one* has actually seen an atom or an electron. What are observed are effects attributed to electrons or atoms. Theoretical concepts that cannot be measured directly in a concrete setting are sometimes called "hypothetical constructs" (MacCorquodale and Meehl, 1948).

Special types of definitions are invented to provide instructions for determining the existence of a theoretical concept in a concrete setting. They are called operational definitions and can be defined as follows:

> Operational Definition—A set of procedures that describes the activities an observer should perform in order to receive sensory impressions (sounds, visual or tactile impressions, etc.) that indicate the existence, or degree of existence, of a theoretical concept.

Operational definitions should be independent of time and space, or abstract, so that they can be used in different concrete settings and at different times.

For example, it may be possible to measure an individual's "anxiety," an emotional state of nervousness or tension, in any one of three ways:

(1) Allow trained observers, i.e., clinical psychologists, to indicate their judgment as to the degree of anxiety;
(2) Collect measures of physiological system activity, such as blood pressure, breathing rate, or sweat gland activity;
(3) Give the individual a questionnaire and study the pattern of responses. (A "yes" answer to the question, "Does the

room feel stuffy?" may be considered an indicator of more anxiety than a "no.")

Each of these procedures may be defined as related to the theoretical concept of anxiety and may be used to measure anxiety in different situations. Each specifies what an individual should do (ignoring certain important technical details) to acquire a sensory impression related to another individual's degree of anxiety (ratings of expert observers, a reading on a dial or chart related to physiological characteristics, or the individual's responses on a questionnaire).

Another example might be the identification of a status hierarchy, or power and prestige rank order, in a face-to-face discussion group. Although it may never be possible to measure the status hierarchy directly, several procedures may be related to the concept of status hierarchy and used to identify it in a particular group, including:

(1) Rank order of the group members in relation to volume of talking (acts initiated);
(2) Rank order of the group members in relation to volume of being talked to (acts received);
(3) Rank order of the group members in terms of their power and prestige in the group as judged by expert observers;
(4) Average rank order of all group members, as perceived by the members of the group, on contributions to or influence over the group task;
(5) Allow members to make decisions about a problem privately and then have the group make a "group decision." Rank the group members according to how close the "group decision" was to their initial decision (the closer their private decision to the group decision, the greater their influence on group decisions and the greater their "power" in the group).

Again, all of these procedures relate to some aspect of the theoretical concept of status hierarchy and instruct the researcher (omitting some details) on how to acquire sensory impressions related to the theoretical concept.

When a theoretical concept is complex, such as the previous example of status hierarchy, many operational definitions may be related to the same theoretical concept, such as the five ways to measure power and prestige rank order. The question arises, should all of the possible operational definitions be positively correlated? Or, should a high reading on one be associated with a high reading on another? In general, the answer is no, for some meaning of a theoretical concept may not be measured by each possible operational definition. For example, rank ordering the members in terms of volume of acts received (talk directed to them) may not be positively correlated with their influence over group decisions. Each operational definition is related to a different aspect of the concept of status hierarchy.

However, here are three kinds of possible relationships between operational definitions. First, there is positive correlation—when one measure is high, the other is high. Second, there is negative correlation—when one measure is high, the other is low. Third, there is no correlation—the value of one measure is independent of another measure; there is no systematic relationship between measures. In general, as long as two operational definitions are not negatively correlated, no problems will arise. This means that they could be positively correlated or not correlated.

For example, consider the concept of "status rank," an individual's position with respect to a single status hierarchy in a small face-to-face group. It might be suggested that there are two ways to measure the "status rank," two operational definitions. First would be the tendency for group members to choose the individual as a desirable social companion, the typical "sociometric" measure. The more one person tends to be chosen by the group members as a desirable social companion, the higher is his "status rank." A second measure would be the group member's perception of the individual as increasing the task effectiveness of the group. The more an individual is perceived as making valuable task contributions, the higher his "status rank."

If these two operational measures of status rank are positively correlated, if high sociometric rank and high task effectiveness rank go together, then no problem exists in identifying an individual's status rank. If these two measures of status rank are not correlated, if sociometric rank seems to have no relationship to task effectiveness rank, then there may be some ambiguity about an individual's overall status rank, but it will not be clear whether this is due to the theoretical concept, status rank, or to the procedures used to measure sociometric rank and task effectiveness rank.

However, if two different operational definitions are negatively correlated, if a person high on the sociometric rank is low on the task effectiveness rank, considerable confusion about the identification of the theoretical concept, status rank, may exist. In other words, if a person is high on the sociometric criterion (he is a desirable social companion) and low on the task effectiveness criterion (he does not contribute much to the group task) what is his overall "status rank"? When this occurs, and there is interobserver agreement on the application of each operational definition (discussed in the next section), steps should be taken to reduce the confusion about the identification of the theoretical concept. This can be done by changing the theoretical concept, resulting in a change in the operational definitions, or by separating the theoretical concept into several theoretical concepts. When Bales found that task rank and social rank were negatively correlated (Henicke and Bales, 1953; Bales and Slater, 1955), he suggested that there were two status hierarchies, a task status hierarchy and a social status hierarchy, thereby dividing the previous concept of a single overall status hierarchy into two separate concepts.

It is clear that the relationship between theoretical concepts and the suitability of operational definitions for measuring the existence of the theoretical concepts is largely one of judgment. Intersubjective agreement about these relationships is low enough without the additional problem of having conflicting measures of the concept, represented by a negative correlation between operational definitions.

A single operational definition can be evaluated by two criteria: (1) its relation to a theoretical concept, and (2) its suitability as a measurement procedure. The previous discussion has focused on the relationship between theoretical concepts and operational definitions and, in particular, on the agreement between two or more operational definitions for measuring a single theoretical concept. The major criterion for evaluating an operational definition as a procedure for measurement is intersubjectivity or interobserver agreement. Will two or more *properly trained* observers arrive at the same results if they independently follow the procedures specified by the operational definition in measuring the same object or phenomenom? High interobserver agreement with respect to a single operational definition means that if two observers *both* measure the same thing, such as sociometric rank, they will both experience the same sense impressions, they will both identify the same person as the sociometric leader. It should be made clear that this is a different issue than agreement between *different* operational definitions, such as task rank and sociometric rank.

When two observers attempt to apply the same operational definition to a phenomenon, there is almost never perfect agreement between observers (random errors alone, errors of measurement, prevent this from occurring). In the actual conduct of research, the issue becomes one of determining what degree of agreement is necessary to answer the research question. The level of agreement that is acceptable varies considerably depending on the research techniques being used and the situation in which they are being applied.

The second criterion, the relation between the operational definition and the theoretical concept, is more difficult to deal with. Careful study of the theoretical concept leads one to consider some measurable characteristic (of a person, group, social system, and the like) that is reasonably related to the theoretical concept. There is no way to avoid the fact that this is largely a matter of judgment. Intersubjective agreement is the only criterion for evaluating the suitability of an operational definition for measuring a theoretical concept. This is the most important step in linking abstract theories to concrete phenomena, and it should be made clear that it is ultimately a matter of scientific judgment.

There is one approach to the development of scientific theory that avoids this problem of subjective judgment. A basic principle of the operationalist position (Bridgman, 1927) is that *all* concepts used in theories should be operational definitions. Although this eliminates the problem of deciding which or how operational definitions are related to theoretical concepts, two additional problems result. First, since there are such a large number of operational definitions possible, and since each new operational definition needs to be incorporated within the theory, the theories become very complex, too complex to be useful. Second, and this is more serious, since all concepts need to be operational definitions, which means they must be directly measurable, it is not possible to include hypothetical constructs, or unmeasurable theoretical concepts, in a theory. This would rule out many concepts that are currently considered useful in both physical and social science (e.g., electrons, status structure), in particular, dispositional concepts (e.g., magnetism, aggressiveness), which refer to the disposition of an object to produce certain effects.

QUANTIFICATION OF CONCEPTS

Previous discussions have treated all concepts equally, no matter what kind of event or phenomenon they describe, such as earthquake, personality, temperature, intelligence, solar system, velocity, small face-to-face group, etc. As it becomes desirable to make more precise statements about events or objects, it becomes necessary to consider different types of concepts. Concepts can be separated into two general groups, those that refer to an object or phenomenon (the name of the concept can be considered a label), and those that refer to characteristics of an object or phenomenon that can differ in degree (the name of the concept is a label for a variety of different situations or states, which can be organized in some systematic fashion). A concept that refers to situations or states that differ in degree is considered to be identical with a concept that differs in quantity—hence the phrase "levels of quantification." When a concept varies in degree, or refers to a phenomenon that assumes different states, each of the states is usually labeled with a number (although for some "levels of quantification," other kinds of labels, such as letters or symbols, would be appropriate).

The quantification of concepts is usually associated with operational definitions, e.g., measuring attitudes on a scale from

−3 to +3; but it can also be applied to theoretical concepts. When applied to operational definitions, the procedures for measurement of abstract concepts in concrete settings, the different types of quantification are generally referred to as "levels of measurement."[4] Four forms of quantification—nominal, ordinal, interval, and ratio—will be examined.

The Nominal Level

If a concept is such that it has only discrete states that bear no obvious relationship to one another, then describing the "states" of the concept amounts to labeling each state of the concept. When the states of a concept can be labeled in *any* fashion, it is considered to be the nominal level of quantification, not to be confused with nominal definitions (definitions composed of primitive terms). A concept with four discrete states may be labeled as follows:

• *A* • *B* • *C* • *D*

or:

• 1 • 2 • 3 • 4

or:

• *G* • *U* • *R* • *K*

or:

• 3.56 • 5 • 10,000 • 43

Any label that will make it easy to refer to each of the four states is suitable.

Many theoretical concepts are conceived on a nominal level of quantification. "Status structure" as a theoretical concept may have only two states: either a group of individuals have one or they do not (such as a collection of people waiting for a bus). Personality types may be considered a nominal concept; individuals may be classified into one of four personality types (i.e., oral, anal, phallic, or genital).

Many operational definitions are on a nominal level. Only a few need be mentioned: sex (male or female)—two states;

4. See Siegel (1956, pp. 21–30) for a good introduction to the subject and Torgenson (1958) for a more sophisticated discussion of different measurement "models," procedures that allow one to examine data and determine the level of measurement appropriate to the data.

marital status (single, married, divorced, or widowed)—four states; zip code—approximately 100,000 states.

In summary, different states of a concept quantified at a nominal level can only be labeled, and no statement about differences between different states is possible, except to say that they are recognized as different.

The Ordinal Level

If a concept is considered to have a number of states that can be rank ordered, it is assumed that some meaning is conveyed by the relative order of the states. For instance, the following rank orders are possible.

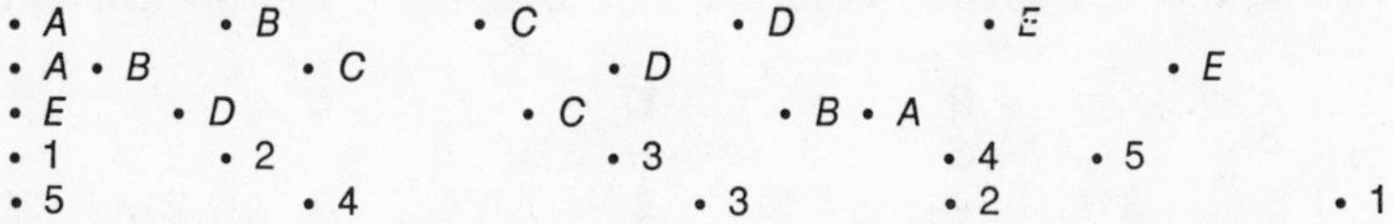

```
• A       • B          • C          • D        • E
• A • B       • C          • D                         • E
• E     • D          • C          • B • A
• 1       • 2              • 3              • 4   • 5
• 5           • 4              • 3          • 2              • 1
```

In each case, it means something for a state to be between two other states, so the labeling procedure must in some way preserve the relative order, or rank order, of the states of the concept; hence the term "ordinal" level of quantification.

The important difference between ordinal and higher levels of quantification is that no information about the difference between states is contained in the labels of the states. The difference between *A* and *B* has no systematic relationship to the difference between *B* and *C*, but *B* is always between *A* and *C*. That is why letters are as appropriate as numbers for labeling ordinal states of a concept, although generally they are not as convenient.

For a student, the most familiar ordinal measures are the grades assigned for academic performance. An *A* always means more than a *B*, and a *B* always means more than a *C*, but the difference between an *A* and a *B* may not always be the same as the difference between a *B* and a *C* in terms of academic achievement.

Most of the theoretical concepts in the social sciences seem to be at an ordinal level of quantification. An individual's relative rank in a small face-to-face group is usually considered an ordinal measure: the individual is first, second, or third in influence. Attitudes are often considered as positive, neutral, or negative. A discussion of the degrees of modernization of a society would probably be in terms of relative rank orders.

In summary, "ordinal level of quantification" applies to concepts that vary in such a way that different states of the concept can be rank ordered with respect to some characteristic.

The Interval Level

An interval level of quantification is one in which states of the concept are not only rank ordered, but also the difference between the states has meaning. Consider the following scales:

• *A*		• *B*		• *C*		• *D*		
• *F*		• *G*		• *H*		• *I*		
• 3	• 3½	• 4	• 4½	• 5	• 5½	• 6	• 6½	• 7
• 45		• 46		• 47		• 48		• 49

In each case, the difference between each adjacent set of points (*A–B*, *G–H*, 4–5, 47–48) is constant. Therefore, comparing the difference between two differences is a meaningful operation. It makes sense to describe *C* as twice as far from *A* as *B*, and *A–C* is twice *A–B.* Since these different states have equal intervals between the states, concepts ordered in this way are said to be labeled on the "interval" level of quantification.

It is important to notice that although the intervals between the states are equal, there is no identifiable zero state. So, although it is reasonable to discuss the difference between 6 and 3 as three times as great as the difference between 6 and 5, state 5 and state 3 cannot be compared directly, since the zero point is not known. If one keeps in mind that the labels are arbitrary, that the same state could be labeled as *B*, *G*, 4, 46, 10.032, or any other symbol, then it is easier to remember that the labels of the states cannot be compared directly. However, numbers are usually used, since "midway between *A* and *B*" is a less convenient description than the symbol "3½."

Although equal-interval theoretical concepts abound in the physical sciences, e.g., temperature, they are harder to find in the social sciences. It might be possible to consider certain types of differences in prestige on an interval scale. Consider three "states" or "levels" of prestige: low, medium, and high. The occupation of fireman might be considered as low, electrical engineer as medium, and physician as high. If the difference in prestige between fireman and engineer is equal to the difference in prestige between engineer and physician, then this theoretical concept of prestige would be quantifiable on an interval scale. If so, physician would *not* have "twice" the amount of prestige as engineer, but the difference between physician and fireman would be twice the difference between physician and engineer.

The classical example of an operational measure on an interval level is that of temperature. It is widely used because most people are familiar with two common scales for measuring temperature, centigrade and Fahrenheit. The difference in temperature, as an abstract concept, between the "state" where water freezes and water boils is a constant interval. But the difference on the centigrade scale is 100 intervals (or degrees) and 180 intervals (or degrees) on the Fahrenheit scale. Two different interval scales are used in operational measures of the same abstract concept, differences in temperature.

Operational measures of social science concepts at the interval level are easier to find than interval abstract concepts. For example, attitudes are frequently measured on a scale like this:

Unfavorable :	____ :	____ :	____ :	____ :	____ :	____ :	____ :	____ :	____ : *Favorable*
	−4	−3	−2	−1	0	+1	+2	+3	+4

If the researcher assumes that the difference between +2 and +4 is the same as the difference between 0 and −2, this is an attempt to apply an interval level of quantification to the measurement procedure.

There is a tendency for social scientists to assume that a measurement has been made at an interval level of quantification even if it is not completely clear that this has been achieved. This is because most of the useful summary measures (such as the "average") and statistical tests require quantification at an interval level. In order to gain the advantages of these statistical procedures, social scientists tend to fudge and act as if the data are at an interval level, particularly if they intuitively feel that the data contain more information than is implied at the ordinal level of quantification.

The Ratio Level

One operation that cannot be carried out with the numbers that indicate locations on an interval scale is that of direct comparisons. For instance, a ratio of two numbers cannot be calculated since the numbers used to label the locations are arbitrary. (The ratio of the boiling point of water to the freezing point is 100:0 using a centigrade temperature scale, and 212:32 using a Fahrenheit temperature scale.) If the location of zero on an interval scale has some meaning, either in terms of a theoretical concept or as a measurement procedure, then it is useful to compute the ratio of two numbers; hence the term, "ratio" level of quantification. It should be mentioned that computation of a ratio is tantamount to comparing two intervals of different lengths that both start at zero.

Some of the best known concepts in physical science are both theoretically and operationally conceptualized at a ratio level of quantification. Time, distance, velocity (a combination of time and distance), mass, and weight are all concepts that have a zero state on an interval scale, both theoretically and operationally. So "twice as far," "twice as fast," and "twice as heavy" have clear and unambiguous meanings. In contrast, one cannot literally speak of "twice as smart," "twice as prejudiced," or "twice the status," since there is no way of knowing where zero smart, zero prejudice, or zero status is.

General Comments on Quantification

In actual research, numbers are used for labels in most operational definitions primarily for convenience. Automated data handling equipment is designed to process numbers; so it is less expensive to record measurements in terms of numbers. Most measurements or operational definitions in the social sciences are at the nominal (labeling), ordinal (rank order), or interval (constant interval, no zero) levels of quantification.

It is much more difficult to quantify theoretical concepts, and many are clearly at a nominal level of conceptualization. For instance, a collection of individuals may have a social structure (if they interact for any period of time) or may not have a social structure (if they never interact). Social structure is clearly conceptualized at a nominal level. However, a characteristic of the social structure, such as the *degree* of status differentiation, may be conceptualized at an ordinal level (rank order) of quantification as high, medium, or low status differentiation.

Another example might be the degree of industrialization of a society. It may be possible to measure a number of "industrial characteristics," such as the number of telephones, volume of hydroelectric power, miles of railroads, and so on, on interval or even ratio scales. Each of these characteristics then becomes one of the operational definitions useful for measuring the theoretical concept, degree of industrialization. However, the theoretical concept itself may not be amenable to more than an ordinal level of quantification. Using these diverse measures (number of telephones, and the like) one may be able to rank order countries *A*, *B*, and *C* in terms of their degree of industrialization but may be unable to compare the "difference between *A* and *B*" with the "difference between *B* and *C*." This problem, developing one measure that includes many diverse and dissimilar characteristics of a single theoretical concept, is called the "problem of aggregation."

One of the major advantages of quantification is that it allows more precise statements to be made about the degree of association, or correlation, between two or more concepts.[5] (This will be treated in more detail in Chap. 4.) One of the important activities in scientific research is discussing the relationships between concepts. Different types of relationship may be categorized into different levels of quantification in the same fashion as concepts. A nominal level of association would be equivalent to the statement that two concepts are or are not related. An ordinal level of association would be equivalent to

5. Two introductory statistical texts (Anderson and Zelditch, 1968; Freeman, 1965) are organized around different measures of association and correlation appropriate for different levels of quantification.

the statement that concepts are positive, not, or negatively correlated. In analyzing data, such statements of association are generally related to operational definitions. They are particularly useful when discussing the interrelations among more than two variables.

One of the most unfortunate aspects of quantification is that it is often mistaken as *the* indication of scientific knowledge. Many individuals firmly believe that if concepts and statements about concepts are not presented in numeric or symbolic terms, then the activity is not really "scientific." The most reasonable response to this attitude is to consider the story of the man who lost his wallet in the middle of the block on a dark night but looked for it at the street corner because the street light was there. Scientific activity should be directed toward asking useful and important questions, even if they are in the middle of the block where the "illumination" provided by quantification is weak. In other words, even an approximate answer to an important question is more useful than a precise, elegant, and quantified answer to a trivial question.

SUMMARY AND CONCLUSION

This chapter can be summarized by the following statement of principles:

(1) If a primitive term is employed, there should be agreement among the relevant scientists about its meaning.
(2) Derived terms should be defined with primitive terms.
(3) There should be shared agreement about the meaning of all theoretical concepts among the relevant scientists.
(4) There should be shared agreement about which operational definitions are indicators of which theoretical concepts.
(5) No two operational definitions should be negatively correlated if they are indicators of the same theoretical concept.
(6) There should be agreement on the measurements obtained when two properly trained individuals apply the same operational definition.

(7) It is not necessary for all theoretical concepts to be measurable in concrete situations.
(8) It is possible to classify operational and theoretical definitions of concepts into four levels of quantification: nominal (labeling), ordinal (rank order), interval (equal interval, no zero), and ratio (equal interval, with zero). Such quantification allows more precise statements about the existence of certain phenomena.
(9) Investigating important scientific questions should not be avoided simply because precise (or quantified) concepts cannot be used.

In conclusion, it should be emphasized that even the most precise, clear, and widely accepted set of definitions, even if they are quantified, can provide no more than a procedure for organizing and classifying objects or phenomena. They cannot be used for prediction, explanation, or providing a sense of understanding. Only when the next step is taken, providing statements that describe relationships between concepts, can these other goals of science be attained. This is an important distinction, for many of the "theories" in social science appear, upon close inspection, to be merely elaborate sets of concepts without any discussion of the relationships between concepts.

4. Statements

Once a concept is presented and there is agreement among scientists about its meaning, it can be used in statements to describe the "real world." This chapter will be devoted to presenting the forms of statements used to express scientific knowledge. Theories, or collections of statements, will be discussed in the next chapter.

Statements can be classified into two groups, those that claim the existence of phenomena referred to by a concept (existence statements) and those that describe a relationship between concepts (relational statements). After a discussion of different types of relations between concepts (such as correlation and causation), different levels of abstraction will be described. Finally, five types of theoretical statements (laws, axioms, propositions, hypotheses, and empirical generalizations) will be examined with respect to their relationship to systematic theory and empirical research.

EXISTENCE STATEMENTS

Statements can be classified into two groups, those that state that a concept exists and those that describe a relationship between concepts. Examples of statements that make existence claims are:

"That object is a chair."
"That chair is brown."
"That object is a person."
"That person has a high authoritarian test score."
"That (object) is a face-to-face group."
"That small face-to-face group has a status hierarchy."

Each of these statements has the same basic form: a concept, identified by a term, is applied to an object or phenomenon. In other words, an object or phenomenon is identified ("object," "chair," "person," "small face-to-face group"), and it is stated that the identified object exists and is an instance of some concept ("chair," "brown," "person," "high authoritarian test score," and so on). Notice that whether the statement refers to existence or degree of existence, e.g., level of authoritarian test score, depends on the level of quantification of the theoretical concept.

Existence statements can be more complex and still retain their basic form. For example:

If (a) there are two or more individuals in group *X*,
(b) each individual can talk to every other individual on a personal basis in group *X*, and
(c) each individual can form a distinct personal impression of every other individual in group *X*,
Then the group is a small face-to-face group.

In this case, a concept, "small face-to-face group," is represented by group *X* because it has the characteristics a, b, and c. The statement claims the existence of an instance of a concept.

Since this form is very similar to the form of a definition, the difference between a definition and an existence statement should be made clear. A definition describes the characteristics of a concept. An existence statement claims that the characteristics and, therefore, instances of the concept actually exist in the "real world." Definitions describe concepts; existence statements claim concepts exist.

Existence statements can be "right" or "wrong," depending on the circumstances. For instance, the statement, "It is noon here," is correct anywhere once a day. The more concrete statement, "It is noon on March 24, 1932, in Paris, France," was only correct once and in one place. In short, differences in levels of abstraction, a distinction to be discussed below, apply to existence statements and affect their potential for "correctness."

RELATIONAL STATEMENTS

There is another type of statement that describes a relationship between two concepts. Knowing the existence of an instance of one concept conveys information about the existence of an instance of another concept. For example:

> If a person is a member of a college fraternity, then he will have a high authoritarian test score.

This statement says that if you identify an individual as a "member of a college fraternity," then you can expect him to have a "high authoritarian test score." Because they describe a relationship between two concepts, statements of this type will be called *relational* statements.

The heart of scientific knowledge is expressed in relational statements. Existence statements, the application of definitions to the real world, can only provide a typology, a classification of objects or phenomena. Explanations, predictions, and a sense of understanding depend on relational statements.

Relational statements can be classified into two broad groups, those that describe an association between two concepts and those that describe a causal relation between two concepts. For instance,

> If a person is a member of a college fraternity, then he will have a high authoritarian test score.

is a statement of association. It says that the person who is in a fraternity will have a high test score. In this context, it does not say that belonging to a fraternity changes a person's degree of authoritarianism. Nor does it say that high test scores will not be found among nonfraternity students. It merely states that the two concepts, fraternity membership and authoritarian test scores, are associated or correlated, and for this reason it is referred to as an "associational" statement.

In contrast, the following statement:

> Membership in a fraternity will increase a person's authoritarian test score.

indicates that one concept, fraternity membership, causes a change in another concept, authoritarian test score. Statements that describe a causal relationship will be called "causal" statements.

Associational Statements

Basically, associational statements describe what concepts occur or exist together. When measures of association at the interval or ratio level of quantification are used, the word "correlation" is often employed to refer to the *degree* of association.

The nature of the association or correlation between two concepts may be considered of three types:

Positive: When one concept occurs, or is high, the other concept occurs, or is high, and vice versa.
Example: Men are taller than women, and vice versa (i.e., taller people tend to be men).

None: The occurrence of one concept gives no information about the occurrence of the other concept, and vice versa.
Example: Male and female students have about the same grades in sociology courses.

Negative: When one concept occurs, or is high, the other concept is low, and vice versa.
Example: Low turnover (changes in membership) in a work group is associated with high productivity, and vice versa.

Notice that a "strong" association (degree of association) between two concepts may occur either positively or negatively, the only difference being the way the concepts are labeled. For example, the following two statements are equivalent, although one expresses a positive association and the other a negative association.

High stability (permanence of membership) in a work group is associated with high productivity. (Positive correlation)

Low turnover (changes in membership) in a work group is associated with high productivity. (Negative correlation)

If it is possible to develop a quantitative operational definition (the result of the measurement operation is a number) for both of the concepts, then it is often possible to represent the degree of association, or degree of correlation, with a number. Most quantitative measures of association are designed to produce numbers from -1.0 to $+1.0$. With such numbers, $+1.0$ indicates a maximum positive correlation, -1.0 indicates a maximum negative correlation, and 0.0 represents no correlation.

Causal Statements

As opposed to relational statements that describe the association or correlation between the occurrence of two concepts, some statements describe the causal relationship between the occurrence of two concepts. In other words, one concept is considered to cause the occurrence of a second concept. For example:

> If a chair belongs to this university, it will be painted brown.
>
> If a small work group has an increase in morale, its productivity will improve.

In each of these situations, one concept is considered to *cause* a state of another concept: belonging to this university causes the chair to be painted brown, and increased group morale causes increased productivity. Statements that describe a causal relationship, sometimes referred to as a cause and effect relationship, will be called "causal" statements. The concept or variable that is the cause is referred to as the "independent variable" (because it varies independently), and the variable that is affected is referred to as the "dependent variable" (because it is dependent on the independent variable).

It is important to point out that, although causal statements and associational statements have very similar forms, often identical in practical use, they are very different types of statements. For instance, it has often been observed that:

> In industrialized societies, the length of women's skirts and the level of economic prosperity are associated; longer skirts are observed during economic recessions and shorter skirts during economic booms.

Despite the empirical support for this statement of association, no one has seriously suggested that there is a causal relationship between these two variables, that longer skirts cause economic recessions, or vice versa. In such a situation, there may be other variables that cause both recessions and longer skirts. This differentiation between the association or correlation between variables and the causal relationship between two variables is often expressed by the principle that *correlation is not (necessarily) causation.*

In practice, determining which relational statements are statements of association and which are statements of causality is often rather difficult, since the form of the statements (the way they are written) is very similar. The reader is forced to make this distinction from the context of the situation. By examining the material surrounding the statement, the reader must infer whether causality or association (correlation) was intended by the writer.

If it is possible to quantify the concepts, or variables, used in causal statements, then it may be possible to describe the degree of causality between the two concepts. In essence, this is the quantification of the concept of causality. Quantification of the degree of causality only makes sense when there is reason to believe that there is more than one independent variable affecting a given dependent variable. For, if there is only one possible cause, then either the proposed independent concept causes the dependent concept or it does not. There is no "middle ground" since there is no other potential cause.

However, if there are two or more independent variables that may have an effect on a single dependent variable, it is reasonable to inquire into the relative influence of these two different causes. This question can only be answered if the dependent variable, the one to be "explained," can be measured at a level above the nominal level (ordinal, interval, or ratio) and preferably above the ordinal level. If the dependent variable can be measured in such a fashion, the researcher usually poses the following question: If a dependent variable can vary (assume different states), how much of this variation is caused by variation in the independent variable(s)?

It is convenient to divide the measured variation (different from the actual variation) of dependent variables into three classes:

(1) That variation directly related to each independent variable under consideration, considering only the variation of that variable;
(2) That variation related to interaction between two or more variables, referred to as the "interaction effect." Strictly speaking, this interaction effect becomes another unique theoretical concept, constituted by the combination of the other concepts;

(3) Variations caused by errors of measurement, unavoidable mistakes in identifying the degree of existence of the dependent variable (the one to be explained).

For example, a researcher may wish to understand the causes of intelligence. He constructs a paper-and-pencil test that an individual can take, and after the researcher scores or grades the test he can assign a score to any individual between 50 and 150, representing his performance in relation to others who took the test. This score is considered a measure of intelligence. The researcher next investigates what causes variation in these scores and decides to investigate two possible causes, parents' intelligence and quality of educational experience. Upon completion of his study, the researcher comes to the following conclusions about the sources (causes) of variance in the intelligence test scores:[1]

SOURCE (CAUSE) OF VARIANCE	PERCENTAGE OF TOTAL VARIANCE (ACCOUNTED FOR) "EXPLAINED"	
	SUBTOTAL	TOTAL
Direct effect of independent variables		50%
Parents' intelligence (high parents' scores associated with high offspring score)	25%	
Quality of educational experience (good schools associated with high score)	25	
Interaction effects		15
High parents' scores and good education (increases score an extra amount)	10	
High parents' scores and poor education	0	
Low parents' scores and good schools	0	
Low parents' scores and poor schools (decreases score an additional amount)	.5	
Errors in measurement (uncontrolled mistakes assumed to occur at random)		25
Unexplained (no answer)		10
Total variance to be explained (always 100%)		100%

This type of summary only identifies the causes of the variance; it does not indicate the exact relationship between the inde-

1. This is an ideal case, used to illustrate the identification of sources of variance, and it is not clear that the current state of methodology would allow such a precise statement to be made on the basis of survey research.

pendent and dependent variables. In other words, the causal process has not been made explicit; only the important independent variables have been described.

Deterministic and Probabilistic Statements

All theoretical statements treated so far in this book have been of this form:

> Under conditions $C_1, \ldots, C_n$, if X occurs, Y will occur.

This is a rather straightforward statement; it says that Y will come after X, period. This type of relation is called "deterministic" because the dependent variable, Y, is *determined* by the independent variable, X.

Another form of relation is possible. It can be described as follows:

> Under conditions $C_1, \ldots, C_n$, if X occurs, Y will occur with probability P.

This is a quite different form of relation, for it indicates that when X occurs Y will occur with probability P and will *not* occur with probability $1 - P$ (all probabilities must sum to 1). Statements containing this type of relation are called *probabilistic.*

Some examples of a probabilistic statement follow:

> In a small discussion group, the emergent leader will be responsible for a given act (spoken comment) with a probability of 0.4.
>
> The probability that a male in the United States will enter the same occupation as his father is 0.10 if his father is (or was) a laborer.
>
> If a die is fair, the probability that any side will come up on a given roll is $\frac{1}{6}$ or 0.1667.

Since the main focus of this book is on the overall development of scientific knowledge, the effects of the differences between deterministic and probabilistic statements will not be treated in detail. But there is no reason to consider probabilistic statements less scientific than deterministic statements as long as the criterion of acceptance is utility in achieving the goals of science.

Because a probabilistic statement predicts both the occurrence and nonoccurrence of an event, the dependent variable, it is impossible to prove it false in any one instance. The usual strategy for testing the usefulness of a probabilistic statement is to study a very large number of events under standard conditions, many more than required by a test of deterministic statements, and compare the empirical results with the prediction of the probabilistic statement. For example, to determine if a die were unbiased (or fair), it might be thrown 10,000 times to see if each side came up approximately 1,667 times (one sixth), as predicted by the statement, "This die is fair."

LEVELS OF ABSTRACTION

Statements can be considered to be on different levels of abstraction, the level of abstraction depending entirely on the level of abstraction of the concepts contained in the statement. It is convenient to consider three levels of abstraction: theoretical, operational, and concrete. The most general is the theoretical level, when a statement contains theoretical concepts. If the theoretical concepts are replaced with the operational definitions related to the theoretical concepts, then the statement is said to be at the operational level. Finally, if the operational definitions are replaced by the findings of a particular research project or the description of a specific concrete event, the statement may be said to be at the concrete level.

It is important to realize that there may be several operational definitions related to each theoretical concept; thus there may be several operational statements related to each theoretical statement. Similarly, there may be several concrete statements related to each operational statement, since each concrete statement is specific to a time and place and operational statements are not. Since there is no real limit on the number of concrete statements, there may be an infinite number of unique concrete statements described in abstract terms by the theoretical statement.

An example of this hierarchy of abstractness can be constructed. Consider this theoretical statement:

> If the rate of succession in an organization is constant, and if the organizational size increases, the degree of formalization will increase.

By replacing the three theoretical concepts (rate of succession, organizational size, and degree of formalization) with corresponding operational definitions (note that the relationship between concepts does not change), the following operational statement is produced:

> If the percentage of employees leaving the firm over a given period of time is constant, and if the number of organizational members increases, the number and explicitness of the organizational rules and procedures will also increase.

Assume that this particular statement was used to guide a research program in a particular organization with the following results, at the concrete level:

> In the *XYZ* organization from June 1, 1961, to May 31, 1965, the percentage of members leaving the firm each month was between 3.8 percent and 4.6 percent (considered to be constant); the organization had 3,000 members on June 1, 1961, and 6,400 members on May 31, 1965, and the organizational manual was found to contain 4,000 words on 200 pages in 600 specific rules on June 1, 1961, and 6,000 words on 325 pages divided into 1,000 specific rules on May 31, 1965.

This statement clearly is subsumed under the operational statement which, in turn, is subsumed under the theoretical statement.

THEORETICAL STATEMENTS

Most "theories" emphasize relational statements, particularly causal statements. But existence statements are often of crucial importance in providing precise descriptions of the conditions under which a relational statement is useful. Consider the following example:

Given:
- (i) A small face-to-face group meets for the first time.
- (ii) The members of the group have different occupations.
- (iii) The members of the group want to do well on a group task,
- (iv) The group task requires the contributions of all the group members.

Then:
- (v) The group members with occupations that are perceived as similar to the group task will have more influence over the group's task activities than group members with occupations dissimilar to the group task.

The first four statements, i–iv, are existence statements; they state the conditions that must exist for statement v to be applied. The fifth statement is a relational statement; it describes a relationship between a characteristic that each group member brings to the situation, his occupation, and a characteristic of his relationship with the other group members, his degree of influence over the group's task activities.

The entire statement may be considered to be of the form:

Given C_1, C_2, C_3, C_4, . . . ; if X, then Y.

C_1, C_2, . . . represent statements i–iv and "if X, then Y" represents statement v. Henceforth, it will be assumed that all theoretical statements will be both abstract and will have this form, a set of conditions describing when a relation between two concepts can be applied. However, it is understood that a number of variations are possible, such as a probabilistic rather than a deterministic relationship between X and Y.

Theoretical statements are frequently referred to by five different labels: laws, axioms, propositions, hypotheses, and empirical generalizations. It is possible for the very same set of words to be all five types of theoretical statement—and simultaneously—depending on the situation. What does differ is the relationship of the theoretical statement to systematic theory and empirical findings. For this reason, these five types of statement will be discussed in relation to these two dimensions, theory and data, in the next two sections.

RELATION OF THEORETICAL STATEMENTS TO THEORY

One of the most familiar terms used in discussing theoretical statements is the word "law." Fundamentally, a law is a statement that describes a relationship in which scientists have so much confidence they consider it an absolute "truth." Three concepts of theory will be discussed in Chapter 5, and one important conception is the view that scientific knowledge is essentially a set of laws, statements that can be considered the "real truth." Most statements that are called "laws" usually contain concepts that can be measured or identified (with the proper operational definitions) in concrete settings.

Another concept of theory, also to be discussed in the next chapter, is the axiomatic form of theory. Axiomatic theory consists of a basic set of statements, each independent of the others (they say different things), from which all other statements of the theory may be logically derived. (High-school plane geometry is usually presented as an axiomatic theory.) The basic statements are known as the "axioms," and the statements that are derived from the axioms are called the "propositions." Some scientists feel that any statement that is used as an axiom in a theory should also be a law, but it is not clear why this should be a necessary characteristic of an axiom, and there is no widespread agreement on this point. The word "proposition" is also used to refer to any idea or hunch that is presented in the form of a scientific statement, similar to the way "hypothesis" is used (described below), but this use has no relation to the propositions of an axiomatic theory.

"Hypothesis" is generally used to refer to a statement selected for comparison against data collected in a concrete situation. The source of the hypothesis may be a variation of a law, a derivation from an axiomatic theory (derived from axioms and propositions), or it may be generated by a scientist's intuition (a hunch); but its most basic feature is that it remains to be compared with empirical data collected in "real life." Because it is to be subjected to empirical test, it is important that all concepts in a hypothesis be measurable, with the proper operational definitions, in concrete situations.

If the same pattern of events is found in a number of different empirical studies, the pattern is often summarized in an "empirical generalization," a "generalization" based on several "empirical" studies. Because they summarize patterns in empirical research, it is obvious that all the concepts in an empirical generalization must be directly measurable. Empirical generalizations are similar to laws, except they do not have as wide acceptance as laws; scientists' confidence in a law is considerably greater than it is in an empirical generalization. Since empirical generalizations represent summaries of research findings, they share no systematic relationship to any particular concept of theory, except as potential laws. However, a theory may be developed to "explain" a relationship summarized in an empirical generalization.

RELATIONSHIP BETWEEN THEORETICAL STATEMENTS AND EMPIRICAL DATA

The next few paragraphs will repeat the same ideas described above, except that a different orientation will be used. Laws, axioms, propositions, hypotheses, and empirical generalizations will be discussed from the point of view of their relationship to data. A thorough familiarity with these five types of statements is so important that this repetition is easily defended if it increases understanding of them. The first idea to be repeated is that sometimes the very same set of words can be viewed as any one of these five types of statement, depending on the scientist's perception of the relationship between the statement and theory or the statement and the empirical data.

The relationship of laws, empirical generalizations, and hypotheses to data is one of degree, not form. Considering the current use of these labels, any concept in a statement called a "law," "empirical generalization," or "hypothesis" must be measurable, using the appropriate operational definitions, in a concrete setting.

Hypotheses are those statements without support from empirical research; it is not yet known whether they are "true" or "false." If one is discussing a research project, it is usually designed to "test" one or more hypotheses. Empirical generalizations, on the other hand, summarize, in general form, the results of several empirical studies, and the scientist usually has some confidence that the same pattern will be repeated in a concrete situation in the future if the same conditions were to recur. Finally, a law is a statement that has so much empirical support that scientists consider it *the* truth. If a research project were to result in data that were inconsistent with *the* truth, a law, scientists would immediately suspect the research project rather than question their confidence in the law.

To summarize, if there is as yet no empirical evidence for or against a statement, it is called a hypothesis; if there is moderate support, it is called an empirical generalization; if the support is "overwhelming," it is called a law. Be warned that since scientists have different standards for evaluating theoretical statements, one man's law may be another man's hypothesis.

The relationship between axioms and propositions, the statements that are part of an axiomatic theory, and empirical data is much less direct. As long as axioms and propositions may contain concepts that are "hypothetical," or cannot be measured directly, their relationships to the empirical world may be more remote than those of hypotheses, empirical generalizations, or laws. However, axioms and propositions may be combined in such a way as to derive statements that contain only empirically measurable concepts, such as hypotheses, in which case their usefulness is determined by the correspondence between these derived statements and the empirical world. This will be discussed in more detail in the next two chapters.

The following statements could be either laws, empirical generalizations, hypotheses, axioms, or propositions, depending on the orientation and purpose of the researcher:

If the volume of a gas is constant, then an increase in temperature will be followed by an increase in pressure.

If the rate of succession in an organization is constant, then an increase in organizational size will be followed by an increase in formalization (of structure and procedures).

If a person has the same attitude toward an object as a good friend, it is unlikely that he will change his attitude toward the object or the friend.

If a person has a different attitude toward an object as a good friend, he may change his attitude toward the object or the friend.

Individuals from upper or elite social classes have more influence in a group composed of individuals from different status classes than individuals from other social classes.

All of these statements have the form of a theoretical statement; they are abstract and describe a relationship between concepts. Because the concepts used in these statements are directly related to operational definitions and therefore can be measured in concrete settings, they can serve as laws, empirical generalizations, or hypotheses.

The following statements could serve as axioms or propositions:

The instincts or needs of the id are satisfied by real or imaginary sense impressions.

If a person disagrees with a friend about his attitude toward an object, then a state of psychological tension is produced.

A state of psychological tension, produced by an imbalanced cognitive structure, produces forces that tend to change the cognitive structure.

However, these statements could *not* serve as laws, empirical generalizations, or hypotheses, simply because they contain concepts that cannot be identified in concrete situations using operational definitions. There is no way to measure the specific instinct activated in the id or the degree of "psychological tension" experienced by an individual. Basically, *any* concept in a statement that is to serve as a law, empirical generalization, or hypothesis must have a direct relationship to concrete situations, utilizing procedures specified by an operational definition.

SUMMARY

Theoretical statements of the form

Given $C_1, C_2, \ldots, C_n$; if X, then Y

are the most important elements of a scientific body of knowledge. They are required before explanations, predictions, or a sense of understanding can be provided. The conditions, C_1, C_2, . . . , C_n, are existence statements that describe the circumstances under which a relational statement—if *X*, then *Y*—can be applied. These statements may be developed at any level of abstraction, but theoretical statements are usually more abstract than operational definitions, which, in turn, are more abstract than descriptions of concrete events.

Relational statements may describe either the association between two or more concepts, as statements of association (or correlation, if quantified), or statements of causality, proposing a causal relationship between two or more concepts.

Theoretical statements may be laws, empirical generalizations, or hypotheses, depending on the degree of empirical support for the statement. And theoretical statements may be laws, axioms, or propositions, depending on their relationship to theory, a systematic set of theoretical statements.

5. Forms of Theories

Scientific knowledge is basically a collection of abstract theoretical statements. At present, there seem to be three different conceptions of how sets of statements should be organized so as to constitute a "theory": (1) set-of-laws, (2) axiomatic, and (3) causal process. The purpose of this chapter is to present examples of these different conceptions of theory and to consider the advantages and disadvantages of each for the purposes of science. One set of statements, developed by Hopkins (1964), will be presented in three different forms, to provide a continuing example.

THE SET-OF-LAWS FORM

One approach is to accept only those statements that can be considered laws as part of scientific knowledge. A set of laws is then considered to be the theory. As mentioned in the previous chapter, all laws are directly supported by empirical research; this means that all concepts used in laws must have operational definitions that allow their identification in concrete situations.

In addition to the potential for measurement of all concepts, some scientists would prefer that only relational statements be called laws, and preferably those that posit a causal relationship rather than an association between concepts. They would rule out statements, such as:

> All social systems have identifiable status hierarchies.

However, this restriction would rule out a number of statements that appear to be useful descriptions of phenomena, such as:

All living things eventually die.

The statement does not state what will cause death, or even when it will occur, but only that the concept "dead" will at some time be an appropriate description of any living thing.

All of the laws described here will contain theoretical concepts that can be measured directly in concrete settings, and most will describe a causal relationship between two concepts. After discussing some examples, the set-of-laws conception of theory will be considered further.

EXAMPLE: *The Iron Law of Oligarchy (Michels, 1959)*[1]

> . . . the majority of human beings, . . . , are predestined by tragic necessity to submit to the dominion of a small minority, and must be content to constitute the pedestal of an oligarchy (Michels, 1959, p. 390).

In more modern terms, using the following concepts:

Social system—Any aggregate of individuals that are organized and have collective interests (i.e., collective goals).

Democratic leadership—Control of all matters affecting a social system or its members, shared *equally* by all the members of the social system, either through direct vote or indirect control over the leaders.

Oligarchical leadership—Complete control of all matters affecting a social system or its members by a privileged clique, drawn from the membership.

the "iron law of oligarchy" now may be restated as follows:

> The only stable form of leadership, even in social systems initially utilizing a democratic leadership, is an oligarchical leadership, and an oligarchy will eventually develop in any social system.

Notice that the "iron law of oligarchy" is in the form of an existence statement. It does not refer to any concept or variable as

1. It is clear that Michels considered this statement as a summary of a set of processes, discussed later in the chapter, and does not suggest that scientific knowledge be organized as a set of laws.

causing the development of an oligarchy; it states only that all social systems will eventually develop an oligarchical form of leadership.

EXAMPLE: *The Laws of Operant Behavior*[2]

The following definitions are important for understanding these laws:

Operant behavior—Any measurable response of an organism, individual, or social system (such as a task-oriented group), that affects or "operates" on its environment, e.g., speech, lever pressing, work activity, etc.

Form of behavior—The actual behavior that is emitted by the organism, e.g., speech patterns may differ, pressing a blue lever is different from pushing a red button, etc.

Rate of behavior—The frequency, in terms of number of actions per unit of time, with which behavior is emitted, e.g., a child saying "daddy" once a minute is emitting behavior at a lower rate than one saying "daddy" ten times a minute.

Reward—Any consequence for the organism that is desirable from the perspective of the organism. Positive rewards increase pleasure and negative rewards reduce pain.

Contingency—The relationship between the past consequences of behavior, its history of rewards from the organism's perspective, and the rate and form of the behavior.

Reinforcement schedule—A particular pattern of contingency, a particular relationship between the occurrence of operant behavior and a pattern of rewards, from the *perspective of an observer* of the organism–environment system. Reinforcement schedules can be classified into two groups:

Continuous reinforcement—A reinforcement schedule in which every operant behavior is followed by the occurrence of a reward.

Intermittent (*partial*) *reinforcement*—A reinforcement schedule in which the occurrence of the rewards is not directly related to behavior in any simple fashion. Rewards might occur after an interval of time (an interval schedule) or after a certain number of behaviors have been emitted (a ratio schedule, referring to the ratio of behaviors to a reward).

2. See Skinner, 1950; Lewis, 1960; Homans, 1961; Hilgard and Bower, 1966; and Burgess and Bushell, 1969, for a summary of these ideas.

FIGURE 5. EFFECT OF CONTINUOUS AND INTERMITTENT REINFORCEMENT SCHEDULES ON THE LEARNING AND EXTINCTION OF OPERANT BEHAVIOR

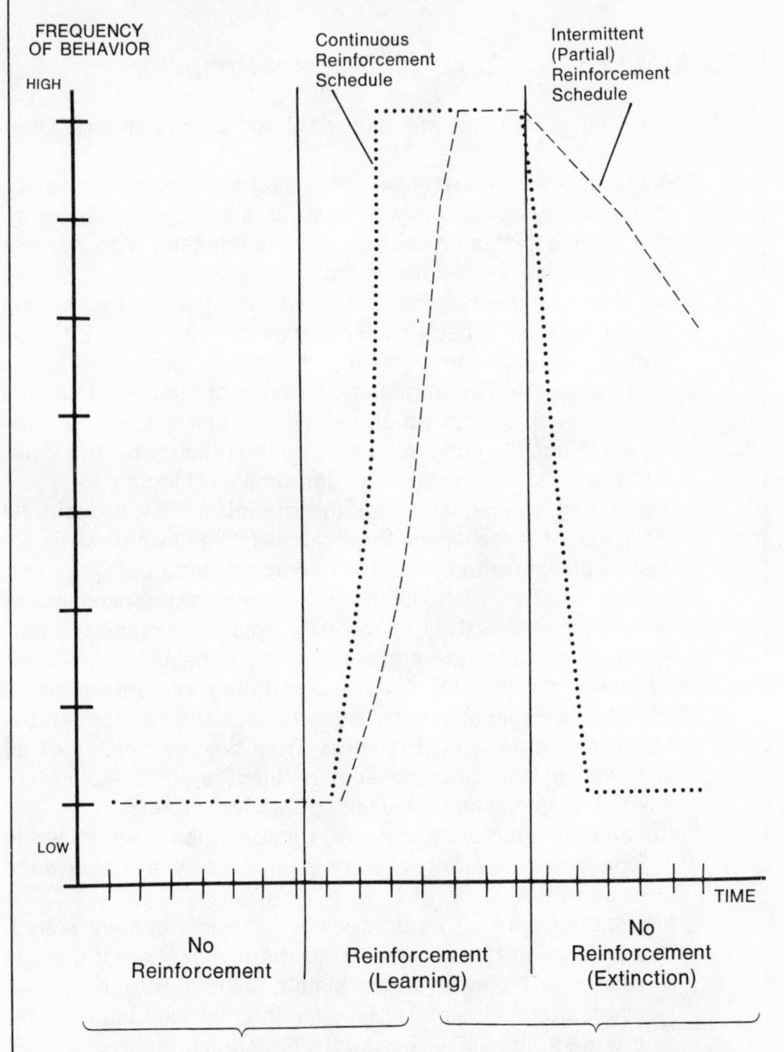

Two types of changes in reinforcement schedules are of interest:

Learning—A change from no reinforcement to any type of reinforcement schedule, continuous or intermittent, from the perspective of an observer of the organism–environment system.

Extinction—A change such that operant behavior previously rewarded, with either a continuous or intermittent reinforcement schedule, is no longer rewarded, from the perspective of an observer of the organism–environment system.

Using these concepts, two laws of operant behavior may be stated:

I. An organism will regularly perform the appropriate (rewarded) behavior sooner in a learning situation if a continuous rather than an intermittent reinforcement schedule is introduced.

II. During extinction, an organism will cease to perform behavior previously rewarded sooner if a continuous rather than an intermittent reinforcement schedule was used before all rewards were terminated.

The patterns of behavior under different reinforcement schedules are presented graphically in Figure 5; these two laws attempt to capture differences between these two patterns.

The first law states that if the relationship between behavior and reward is continuous, or certain, then the organism will emit the rewarded behavior more quickly than if the relationship is intermittent (problematic). The second law says that if rewards terminate, an organism accustomed to a continuous reinforcement schedule will terminate behavior sooner than an organism accustomed to an intermittent schedule.

One way to consider the development of a theory using the set-of-laws conception of theory is to consider abstract theoretical statements as having different degrees of empirical support, as discussed in Chapter 4. Those with no support are considered hypotheses, those with some support are considered empirical generalizations, and those with "overwhelming" support are considered laws. In the following example, a set of statements is organized in this fashion. This is possible because the theorist, Hopkins (1964), carefully examined the degree of empirical support for each statement he included in his theory.

EXAMPLE: *The Exercise of Influence in Small Groups (Hopkins, 1964)*

The following concepts, all referring to the characteristics of the members in a small face-to-face group, are used by Hopkins:

Rank—The generally agreed upon worth or standing of a member relative to the other members, as evaluated by the group members.

Centrality—Closeness to the "center" of the group's interaction network; thus refers simultaneously to the frequency with which a member participates in interaction with other group members and the range of other group members he interacts with.

Observability—Relative ability to observe the actual norms of the other group members and hence the norms of the group.

Conformity—Degree of congruence between a member's actual belief in relation to a norm and the group position on that norm (the average of the other members).

Influence—Relative influence of the member on the actions of the other group members.

In discussing his theory, Hopkins (1964, pp. 97–98) considers the relative amount of empirical support for each of the fifteen propositions in his theory.[3] These statements, classified in relation to the amount of empirical support they have received, in Hopkins's judgment, are:

(1) *Good* empirical support, status of laws.
If rank, then centrality.
If observability, then conformity.
If centrality, then influence.
If centrality, then rank.

(2) *Some* empirical support, status of empirical generalizations.
If centrality, then observability.
If rank, then observability.
If centrality, then conformity.
If rank, then conformity.
If conformity, then observability.
If influence, then conformity.
If rank, then influence.
If influence, then rank.

3. Hopkins presents his ideas in axiomatic form; this interpretation is used only to illustrate the set-of-laws conception of theory. The material from Terence K. Hopkins, *The Exercise of Influence in Small Groups* (Totowa, N.J.: The Bedminster Press, 1964), is used with the permission of The Bedminster Press and the author.

(3) *No* empirical support, status of hypotheses.
If observability, then influence.
If conformity, then influence.
If influence, then observability.

With these examples, it is possible to consider the usefulness of a set-of-laws theory for achieving the goals of science. First, the concepts in these statements can be used to classify and organize (provide a typology for) the phenomenon under study. Social systems may be classified according to whether they have a democratic or oligarchical form of leadership. An organism's rate of operant behavior may be classified as either high or low. The relationship between the behavior and the consequences to an organism, a schedule of reinforcement, may be classified as continuous or intermittent. An individual in a small face-to-face group may be classified in terms of his rank, centrality, observability, conformity, or influence. In every case, classification is related to the concepts of the law, or set of laws, and is consistent with the conception of a theory as a set of laws. The first purpose of scientific knowledge, providing a typology, is clearly achieved with a set-of-laws form of theory.

The next purpose of scientific knowledge, providing logical explanations and predictions, can also be achieved using a set of laws. Each of the three examples will be used to predict or explain some form of concrete behavior.

(i) At one point in time, say year *X*, a small country, Gamma, has a political system in which each citizen not only has the right to vote on all important issues, but he is fined for failure to exercise his privilege. Some years later, $X + 2{,}000$ years, it is observed that one small group of individuals is making all the decisions for the citizens of Gamma, even those related to matters of personal attire and hair style. Clearly, Gamma had a democratic form of government in year *X* and an oligarchical form of government in year $X + 2{,}000$. Michels's "iron law of oligarchy," which states that in any social system a democratic form of leadership will change to an oligarchical form of leadership, can be used to logically predict or explain this change in the political system of Gamma.

(ii) Two students attend a large university and study in the library, taking occasional breaks for a drink. One student, named Joe, uses the student lounge in the basement where an unpredictable drink machine works about half the time. The other student, Harry, uses the student lounge on the top floor where the drink machine always works properly. Strictly by chance, both drink machines become inoperative on the same day. Harry, who had been using the "good" machine, stops putting money in the machine after he loses his first dime while Joe, who had been using the "unpredictable" machine, continues to put dimes in the machine in an attempt to acquire a drink.

These differences in behavior can be explained by operant behavior law II. Harry's exchanges with the "good" machine can be considered a continuous reinforcement schedule and Joe's exchanges with the "unpredictable" machine can be considered an intermittent reinforcement schedule. Operant behavior law II states that an individual, such as Joe, accustomed to an intermittent reinforcement schedule will emit behavior (put dimes in the machine) more frequently after rewards (the drinks) are discontinued than those, such as Harry, accustomed to a continuous reinforcement schedule.

(iii) Finally, consider one of the four "laws" in the formulation of the "small group influence theory" developed by Hopkins (1964): "If centrality, then rank." This can be used to predict the relative rank, or prestige, of individuals that vary in terms of their centrality, or closeness to the center of the interaction network, in a face-to-face group. For instance:

Law: In a small face-to-face group, if centrality, then rank.

Situation *X*: In a small work group, Harry is involved in more group interaction than any other member; he has high centrality.

Therefore: Harry should have high rank (prestige) in the group.

The other three laws may be used to predict or explain other characteristics of Harry's relative standing in the group.

Scientific knowledge in the form of a set of laws appears to be useful for providing a typology, providing predictions and explanations, and, if the statements are sufficiently precise, allowing the potential for control. However, they do not provide any "sense of understanding" with regard to any of the phenomena discussed. What social processes cause all organized collectivities to develop oligarchical leadership structures? What mental processes cause individuals on intermittent schedules to take longer to learn that certain behavior is rewarding and longer to learn that certain behavior is no longer rewarding, compared to those on continuous schedules? What group processes cause individuals more central in an interaction pattern to acquire higher prestige (rank) in a face-to-face group? These types of questions cannot be answered by a theory in the form of a set of laws. Therefore, if scientific knowledge is organized in the form of a set of laws, a scientist cannot achieve all the purposes of science, since he cannot provide a sense of understanding.

This conception of scientific knowledge has several other disadvantages. First, since laws are directly supported by empirical research, every concept used in a law must have at least one operational definition that allows identification of the theoretical concept in concrete settings. This prohibits the use of any unmeasurable concepts or hypothetical constructs in theoretical statements. As mentioned earlier, this would prohibit the use of many concepts currently employed in science, particularly dispositional concepts, which refer to the tendency of "things" to create certain effects, i.e., magnetism, authoritarianism, etc. One can only measure the consequences of a dispositional concept, i.e., attraction to iron, tendency to perceive in absolutes (good or bad), and the like, but not the actual concept itself.

Second, the statements that compose a set of laws are supposed to be independent, unrelated to one another. This has several disadvantages. It means that the final set of statements will be very large, since the relationship between every set of concepts requires a theoretical statement or law. Also, since the statements are considered to be independent, research in support of one statement or law cannot provide support for another statement or law. Compared with other ways of organizing theoretical statements, the set-of-laws form may require more research and therefore may be comparatively inefficient. However, this depends on which research strategy is being employed. (For a fuller discussion, see Chap. 7.)

In summary, the conception of scientific knowledge as a set of laws will allow scientists to achieve some of the goals of science—typologies, prediction and explanation, and if the laws are carefully specified, the potential for control. However, a sense of understanding is completely absent when laws are used to logically explain phenomena. The attempt to concentrate on developing a set of laws eliminates unmeasurable or hypothetical theoretical concepts from the repertoire of available concepts, results in a very large set of statements (that may be difficult to organize in a useful form) and a much larger program of research, since each law must receive empirical support independent of empirical support provided for any other law.

As a final comment, most of human knowledge, perhaps more than 90 percent, consists of statements that meet the criterion of laws or empirical generalizations, that is, a relationship between two observed concepts. This is particularly true of "practical knowledge," such as medicine, engineering, or administration, where "rules of thumb" are employed, often disguised as "judgment," without precise knowledge of the specific causal processes involved. On the other hand, the lack of theories to explain this practical knowledge, including a sense of understanding, suggests that there is much work left for scientists in these fields.

THE AXIOMATIC FORM

An axiomatic theory is typically defined as an interrelated set of definitions and statements with several important features:

(1) A set of definitions, including theoretical concepts, both primitive and derived (nominal), and operational definitions (to allow the identification of some abstract theoretical concepts in concrete settings).

(2) A set of existence statements that describe the situations in which the theory can be applied, sometimes referred to as the *scope conditions* since they describe the scope of conditions to which the theory is considered applicable. (These statements are not required in a completely imaginary theory, such as in mathematics, that is not intended to be applied to concrete or "real" phenomena.)

(3) A set of relational statements, divided into two groups:

(a) Axioms—A set of statements from which all other statements in the theory may be derived.

(b) Propositions—All other statements in the theory, all derived from combinations of axioms, axioms and propositions, or other propositions.

(4) A logical system used to:
   (a) Relate all concepts within statements, and
   (b) Derive propositions from axioms, combinations of axioms and propositions, or other propositions.

Plane geometry is perhaps the most widely known axiomatic theory; in fact, most mathematical theories are in this form.

It is extremely difficult to find examples of axiomatic theory relating to social or human phenomena, and it is even harder to find those that are simple enough for an introductory discussion. However, Hopkins (1964) does present his statements in this form, and his material will serve as an example.

EXAMPLE: *The Exercise of Influence in Small Groups (Hopkins, 1964)*

The crucial definitions in this theory are those of rank, centrality, influence, observability, and conformity, all defined in the previous section.

The only scope condition is that the theory applies only to small interacting groups, where each member has the opportunity to form a personal impression of every other member.

Hopkins selected nine statements as axioms:

A–1	If rank, then centrality.
A–2	If centrality, then observability.
A–3	If centrality, then conformity.
A–4	If observability, then conformity.
A–5	If conformity, then observability.
A–6	If observability, then influence.
A–7	If conformity, then influence.
A–8	If influence, then conformity.
A–9	If influence, then rank.

These axioms can be combined to produce new statements, or propositions. For example, axioms A–1 and A–2 may be combined to produce a new proposition as follows:

A–1	If rank, then centrality.
A–2	If centrality, then observability.
Therefore:	If rank, then observability.

Another set of axioms may be combined to produce the same proposition.

A–1	If rank, then centrality.
A–3	If centrality, then conformity.
A–5	If conformity, then observability.
Therefore:	If rank, then observability.

Using all possible combinations of axioms, it is possible to produce eleven propositions, some in as many as four ways (four different combinations of axioms). These propositions and the axioms used in their derivation are presented below:

Axioms Used in Derivations	*Propositions*	
A–1, A–2 A–1, A–3, A–5	P–1	Rank, then observability.
A–1, A–3 A–1, A–2, A–4	P–2	Rank, then conformity.
A–2, A–6 A–2, A–4, A–7 A–3, A–7 A–3, A–5, A–6	P–3	Centrality, then influence.
A–8, A–7 A–9, A–1, A–2	P–4	Influence, then observability.
A–1, A–3, A–7 A–1, A–3, A–5, A–6 A–1, A–2, A–6 A–1, A–2, A–4, A–7	P–5	Rank, then influence.
A–3, A–7, A–9 A–3, A–5, A–6, A–9 A–2, A–6, A–9 A–2, A–4, A–7, A–9	P–6	Centrality, then rank.
A–9, A–1	P–7	Influence, then centrality.
A–7, A–9 A–5, A–6, A–9	P–8	Conformity, then rank.
A–6, A–9 A–4, A–7, A–9	P–9	Observability, then rank.
A–7, A–9, A–1 A–5, A–6, A–9, A–1	P–10	Conformity, then centrality.
A–6, A–9, A–1 A–4, A–7, A–9, A–1	P–11	Observability, then centrality.

This list is longer than the original list in the previous section because Hopkins did not consider propositions P–7 through P–11 "theoretically interesting" and did not discuss them.

One of the most important problems in dealing with theories in axiomatic form is determining how to select the axioms. In other words, what criteria will be used to choose certain statements and call them axioms. Consistency is clearly required; no two axioms, or any combination of axioms, should make conflicting predictions. But, aside from the criterion of consistency, it is difficult to establish criteria for substantive theories.

In dealing with logical systems that are completely abstract and independent of any feature of the "real world," as in mathematics, a common criterion is to select the smallest set of axioms from which all other statements can be derived, reflecting a preference for simplicity and elegance. There is reason to think that this is inappropriate for a substantive theory, particularly when it makes it more difficult to understand the theory.

One criterion that is seriously entertained is to accept as axioms only those statements that have achieved the status of laws.[4] If this criterion is accepted, an axiomatic form of theory becomes a procedure for organizing and integrating a set of laws, which has merit. However, it will be remembered that any statement that becomes a law must have considerable empirical support *before* it is considered a law. This would mean that all the statements and the concepts contained in the statements used as axioms must have a direct correspondence to concrete settings. This would prevent the inclusion of hypothetical or unmeasurable concepts in an axiomatic theory, thereby eliminating one of its most useful advantages.

In dealing with axiomatic theory related to substantive matters, it seems appropriate to select as axioms that set of independent statements that makes the theory easiest to understand, no matter how large. Blalock (1969, p. 18) suggests that this is achieved if only those statements that describe a *direct causal* relationship between two concepts are employed as axioms. This is the criterion used by Hopkins (1964), and it seems to be a reasonable suggestion. However, if it is found that some other set of statements is more clearly understood (there is more shared agreement among the audience), the

4. Costner and Leik (1964) assume this criterion in their critique of the problems of testing theory in axiomatic form.

theorist should feel free to use it. This is consistent with Blalock's goal, which is to reduce the ambiguity in the description of theories.

As with the statements in the set-of-laws concept of theory, the statements in the axiomatic form of theory can be used to logically derive explanations and predictions. The concepts contained in the theory can also be used to classify and organize events. But, since the logical notion of explanation is utilized with this form of theory, it fails to provide a "sense of understanding."

However, the axiomatic form does have several advantages over the set-of-laws form of theory. First, since some statements can be derived from others, it is not necessary for all concepts to be measurable. Hence, unmeasurable or hypothetical concepts may be employed in developing the theory. Second, the number of statements that express scientific knowledge can be smaller. Instead of requiring a statement to describe a relationship between every pair of concepts, a set of axioms and the logical system may be used to generate a large number of statements. The set of axioms and the logical system may be less cumbersome than a large set of independent statements. Third, research may be more efficient. Because the theory is an interrelated set of statements, empirical support for any one statement tends to provide support for the entire theory and, hence, support for the other statements that compose the theory. This form of theory is more appropriate for the theory-then-research strategy, discussed in Chapter 7. Fourth, the axiomatic form allows the theorist to examine *all* the consequences of his assumptions, or axioms. Often the careful formalization of a theory results in some surprises, or "unintended consequences," that derive from a certain conceptualization or paradigm.

Finally, the axiomatic form of theory is compatible with the causal process form of theory described in the next section. Frequently, the statements that compose an axiomatic theory can be organized in such a way as to provide a causal description of the process that relates changes in an independent variable with changes in a dependent variable. If this is done, they can provide a "sense of understanding." However, this is not always possible.

The lack of social science theories in axiomatic form suggests that it has either been impossible or inconvenient for social scientists to put their ideas into this format. However, almost all of the paradigms in social science appear amiable to the following form of theory, the causal process form.

THE CAUSAL PROCESS FORM

The causal process form of theory is an interrelated set of definitions and statements with the following features:[5]

(1) A set of definitions, including those of theoretical concepts, using both primitive and derived (nominal) terms, and operational definitions (that describe how to identify some of the theoretical concepts in concrete settings).
(2) A set of existence statements that describe those situations in which one or more of the causal processes are expected to occur, or as it is sometimes described, when these processes will be "activated."
(3) A set of causal statements, with either deterministic or probabilistic relations, that describe one or more causal processes or causal mechanisms that identify the effect of one or more independent variables on one or more dependent variables. Although different causal mechanisms may differ in impact on the dependent variables, all statements are considered of equal importance in terms of the presentation of the theory.

The major difference between this form of theory and the axiomatic form is that all statements are considered to be of equal importance, they are not classified into axioms and propositions, and the statements are presented in a different fashion, as a causal process. Intersubjective theoretical concepts are still required, hypothetical (unmeasurable) concepts are al-

5. Berger *et al.* (1962) discuss theoretical-construct models (pp. 67–101), which are identical to the causal process model discussed here. Blalock (1969) discusses causal models and puts them in the form of systems of linear equations. Both discussions are more sophisticated, and complex, than the discussion here.

lowed, and scope conditions, describing when and where the causal process will occur, are necessary.

No matter what type of theory a scientist claims to be dealing with, when he explains "how" something happens he usually refers to a description of one or more causal processes. Therefore, it seems reasonable to openly acknowledge the description of a causal process as a theoretical form and consider its advantages. A further discussion of the causal process form follows these examples.

EXAMPLE: *The Effect of First Impressions on Cognitions*

The following statement, taken from Secord and Backman (1964, p. 59) can be considered an empirical generalization:

> When an individual forms an impression of another person, information about the other received earlier has more effect on the final impression than information about the other received later.

One causal process that will explain this finding is composed of the following statements:

Scope Condition: This process is expected to be activated at any time the individual forms, for the first time, an impression of a complex object, including other individuals.

Under this condition, the following causal process will be activated:

(i) When a person develops a cognition of a complex object, an overall framework is developed before the details are completed.

(ii) Information received first is used to develop the overall framework of the new cognition.

(iii) Information received later is used to complete the details of the cognition.

(iv) Final impressions reflect the general outlines of the new cognition more than they reflect the details.

This process can be summarized by stating that initial information is used to develop a general outline that is refined as more details are acquired. However, the overall impression is a reflection of the general outlines, which are not significantly affected by the details. Hence, the overall impressions are influenced more by the initial information than by later information.

The remaining examples will not be presented in statement form, but they could be if it were necessary. Sophisticated and precise theory construction is enhanced by using such an explicit technique to trace the causal linkages in a theory. These examples are presented to provide a general notion of the causal process form of theory as compared with other forms.

EXAMPLE: *Creation of Oligarchies (Michels, 1959)*

The "iron law of oligarchy" is:

> The only stable form of leadership, even in social systems initially utilizing a democratic leadership, is an oligarchical leadership, and an oligarchy will eventually develop in any social system.

Michels was studying European workers' organizations in the late nineteenth century when he wrote his book. These are a combination of unions and political parties, as they are known in the United States. Michels suggested that four processes contributed to the development of oligarchical leadership in "political parties."

The first and most fundamental of these processes is the "technical necessity" for leadership in any large organization. Any collectivity that numbers in the tens of thousands, as did the workers' organizations, requires some coordination and organization in order to establish and achieve *any* goals. Members from the ranks "emerge" as leaders to carry out this function, coordinating and organizing the "party." Those members selected to act as leaders soon achieve power, prestige, and often income greater than that of the ordinary party member. However, since in the initial stages of organizing the party the goals of the members and the leaders are identical—achievement of the party goals—and since the leaders have recently been selected from the workers by the workers, the leadership reflects the desires of the members and can be considered democratic.

Once the initial stages of organization, emergence of leaders, and the achievement of initial party goals are past, Michels claims that three processes are activated that contribute to the development of an oligarchy, a leadership insensitive to the desires of the members. The first of these is the unwillingness of the new leaders to relinquish their positions of higher status. Having enjoyed prestige, income, and the exercise of power, they are reluctant to allow themselves to be replaced or return to their former status as ordinary members. In short, they fight like hell to keep on top, even if it requires co-opting potential rivals or engaging in ruthless and unethical activity (i.e., threats, violence, or murder) to prevent rivals from taking over.

The second of these processes is the general tendency of the members to be somewhat awed by the high status of the leaders and grateful for their help in achieving the party goals. Individually, the members tend to be satisfied with the leaders as long as the collective goals are achieved, and they show little interest in the actual leadership activities.

The third of these processes is the general tendency of the members of political parties to be generally disinterested in party activities on a day-to-day basis. In other words, the demands of job, family, friends, church, and so on, take precedence over the party as long as no major problem arises.

Michels suggests that these three processes, the tenacity and ruthlessness of the leaders in maintaining their positions, the tendency of members to be grateful with respect to the leaders, and the general disinterest of the workers belonging to a "successful" party in its activities, result in a stable clique of individuals performing the functions of leadership, often independently of the interests of the apathetic party members. If this continues until the leaders have control of any mechanism the members might use to influence their decisions (such as the election of new leaders), then the leadership structure has changed from democratic to oligarchical, and the members have become "the pedestal of an oligarchy."

Michels then suggests that if the political parties have an oligarchical leadership structure, then the society will have an oligarchical political structure.

Although Michels's explanations of processes have not been presented formally, they are rather straightforward and provide a sense of understanding that is clearly absent from the "iron law of oligarchy" when it is presented without the processes that explain it.

EXAMPLE: *Operant Behavior, Law II*

During extinction, an organism will cease to perform behavior previously rewarded sooner if a continuous rather than an intermittent reinforcement schedule was used before all rewards were terminated.

Describing a cognitive or mental process that will explain this law is rather straightforward if it is assumed that any organism will develop ideas about the relationship between its behavior and the occurrence of rewards. Explaining the law only requires that one speculate on the nature of the "theory" that organisms will develop about the relation between their behavior and the rewards.

Consider an organism in a rewarding situation where the reinforcement schedule is continuous—every time a behavior occurs, a reward occurs. The organism may well develop a very simple and straightforward view of his relation with his environment—the proper behavior (i.e., pressing a key, doing homework properly) is promptly rewarded. If, from the point of view of one observing the organism–environment system, a condition of extinction occurs—the rewards stop coming—we can consider the cognitive processes of the organism. Having been accustomed to a reward for each behavior, the organism may quickly notice that the appropriate behavior is not producing a reward. After a few additional behaviors, as a final check, the organism stops the behavior, confident that the environment has changed. In operant-conditioning jargon, extinction of the behavior has occurred.

Consider an organism that has been receiving rewards on an intermittent reinforcement schedule. This organism emits behaviors and every once in a while receives a reward. Assuming that the organism is trying to develop a private "theory" about the relationship between the behavior and the rewards, it is unlikely that it will either be a very simple theory or one the organism trusts very much, since there may actually be no straightforward relationship between the rewards and the behavior. When the extinction occurs (an observer knows that no further rewards will occur), how can the organism determine that no further reward will occur? Since the organism may have had no thorough understanding of the relationship during reinforcement, it may be very difficult for him to test the notion that rewards will no longer appear. The organism may try out a number of variations, including long sequences of similar behavior, in an attempt to exhaust all combinations of behavior that will produce rewards. As a result, behavior is emitted long after the reinforcement schedule is terminated, and it takes longer to extinguish the behavior. As Skinner suggests (1950, p. 205), "the situation (for the organism) in extinction is not wholly novel (or new)."

In summary, two variations of mental processes are suggested to explain the law:

Continuous Reinforcement→No Reinforcement:
Organism develops a simple "theory" about the relationship between his behavior and the rewards. When the rewards stop, the organism can easily test the "theory" and assumes that there will be no more rewards. Behavior stops soon after rewards stop.

Intermittent Reinforcement→No Reinforcement:
Organism attempts to develop a "theory" of rewards, but the complexity of the relationship prevents any simple or unambiguous "theory" from being established. When the rewards stop coming, the organism must choose between the conclusion that his "theory" is wrong or that the situation has changed. Since the theory is complex and ambiguous, it takes longer to test all possible alternatives, and the behavior continues long after the rewards stop.

EXAMPLE: *The Exercise of Influence in Small Groups (Hopkins, 1964)*

A number of processes were described in the original version of this theory. However, Hopkins suggested that they were all activated in certain situations, encompassed by the following scope statement:

In any interacting group, where each member has the opportunity to form a personal impression of every other member, the processes related to the exercise of influence are activated.

One statement, with enough empirical support that it can be considered a law, has been chosen for this example:

"If centrality, then rank."

The following set of statements describe a causal process that will explain this statement:[5]

5. The original formulation was a completely circular process. It is presented in this form, broken at the "If rank, then centrality" link to provide a suitable example.

If centrality, then conformity.
If centrality, then observability.
If conformity, then observability.
If observability, then conformity.
If conformity, then influence.
If observability, then influence.
If influence, then rank.

They can be represented diagrammatically as a causal process as follows:

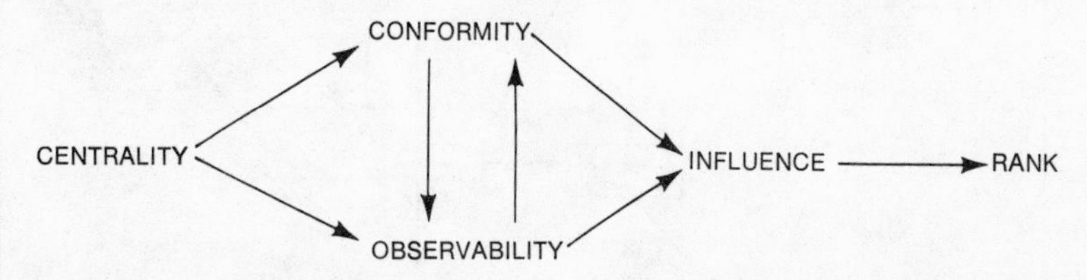

Given this process, a number of additional statements may be derived:

If centrality, then influence.
If conformity, then rank.
If observability, then rank.

The following example is presented to demonstrate another level of sophistication that is possible with the causal process concept of theory. Notice that, although the formulation is rather elaborate, particularly for social science, this model of theory makes the relationships between concepts relatively clear. It would be easy to do research related to this theory without becoming confused about the relationship between the theoretical statements, although the methodological problems may become quite complex.

EXAMPLE: *Status Incongruence and Mental Health*

This theory by Kasl and Cobb (1967) requires understanding of the concept of "status stress," which is defined as a situation in which an individual has different ranks on two or more societal status hierarchies, e.g., an individual with much education and little income, or vice versa. (See the discussion of status inconsistency in Chap. 2, p. 31.)

Figure 6 shows the causal processes that constitute this theory; it can be used to derive the following statements:

FIGURE 6. A THEORETICAL FORMULATION OF THE EFFECTS OF PARENTAL STATUS INCONGRUENCE AND DISCREPANCY ON THE PHYSICAL AND MENTAL HEALTH OF ADULT OFFSPRING

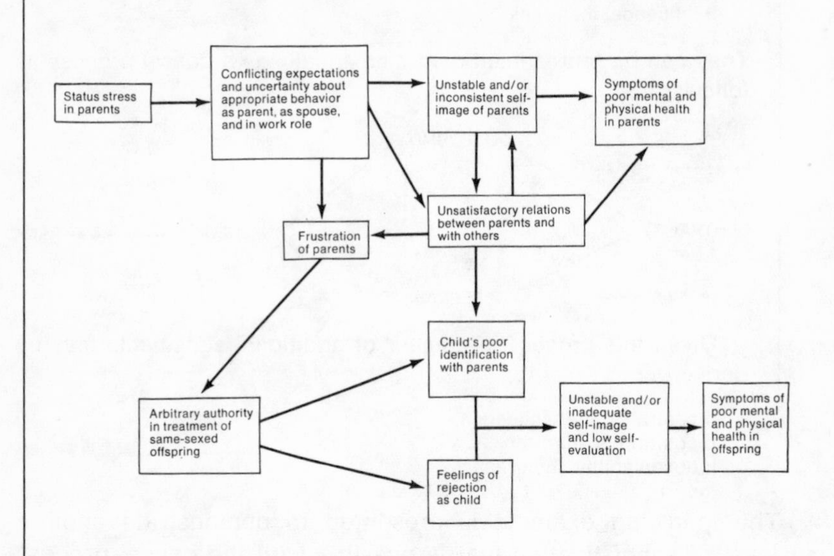

Taken from Stanislav V. Kasl and Sidney Cobb, "Effects of Parental Status Incongruence and Discrepancy on Physical and Mental Health of Adult Offspring," *Journal of Personality and Social Psychology* (October 1967), 7(2), Part 2, Whole No. 642:1–15. Reprinted by permission of the American Psychological Association and the authors. All arrows represent positive correlations.

(i) Status stress in individuals produces symptoms of poor physical and mental health in the individuals.

(ii) Status stress in parents produces symptoms of poor physical and mental health in offspring.

With this conception of theory, there is a major problem of definition. What is the boundary of the causal process? The example of a causal process taken from Hopkins (1964) may be represented diagrammatically as follows:

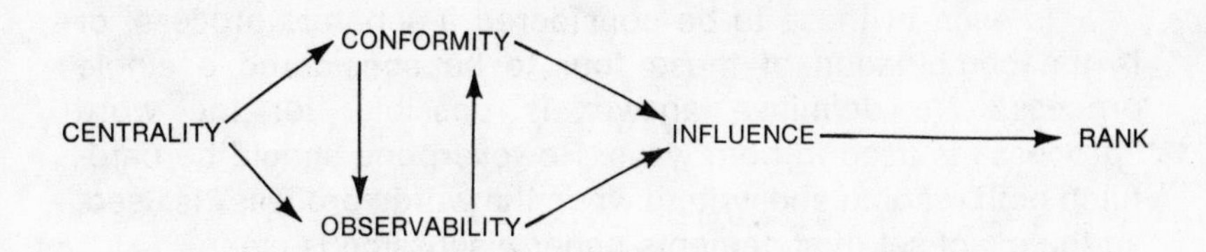

Or the process may be presented as *four* different causal processes, as follows:

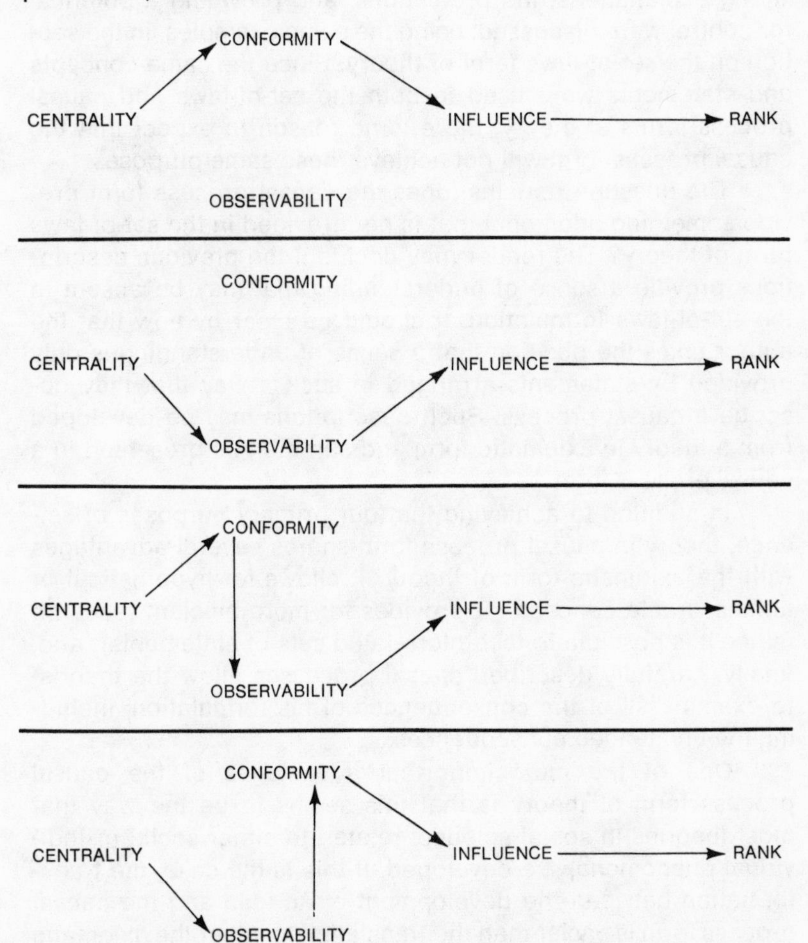

Is each of these to be considered a separate process, or is the combination of these four to be considered a single process? No definitive answer is possible, for the word "process" is used in both ways. However, one should be careful in both reading and writing when the word "process" is used, so the exact set of statements under discussion is clear.

Is the causal process form of theory useful for the purposes of science? Three of the purposes, providing a typology, providing explanations and predictions, and providing a potential for control were discussed, using the same examples, in the section on the set-of-laws form of theory. Since the same concepts and statements were used in both the set-of-laws and causal process forms of theory, there is no reason to expect that the causal process form will not achieve these same purposes.

The question remains, does the causal process form provide something additional that is not provided in the set-of-laws form of theory? The reader may decide if the previous descriptions provide a sense of understanding that may be absent in the set-of-laws formulation. It should be clear by now that the author takes the position that a sense of understanding is only provided by statements arranged in such a way that they describe a causal process. Such descriptions may be developed from a theory in axiomatic form and can then be presented in a causal process form.

In addition to achieving the four primary purposes of science, theory in causal process form shares several advantages with the axiomatic form of theory: it allows for hypothetical or unmeasurable concepts; it provides for more efficient research (since it is possible to test interrelated sets of statements); and, finally, carefully described causal processes allow the theorist to examine all of the consequences of his formulation, including the unintended consequences.

One of the most important advantages of the causal process form of theory is that this seems to be the way that most theories in social science, related to either social or individual phenomena, are developed. If this is the case, the transformation between the development of an idea and the causal process form is easier than the translation to either the axiomatic or set-of-laws form of theory.

One problem exists with the causal process form of theory: When do you stop? In other words, at what point can a theorist decide when all the steps, or statements, in a causal linkage have been specified? There seems to be no objective answer to this question, and perhaps the only solution is to require intersubjective agreement among the relevant scientists. Only when a researcher and his colleagues agree that all the steps in a causal process have been identified is it time to stop working on the theory.

EVALUATION OF THE THREE FORMS OF THEORY

It was concluded earlier that any of three forms of theory are suitable for three purposes of science: providing a typology, logical explanation and prediction, and potential for control. But *only* the causal process form, or statements from an axiomatic theory put in causal process form, can provide a sense of understanding.

While often statements from an axiomatic form of theory can be placed in a causal process form, it is not always the case. Several statements from an axiomatic theory may be combined to produce a logical prediction for which no reasonable causal process exists. Nevertheless, for the purposes of the following discussion, they will be treated jointly and referred to as the "axiomatic-causal process."

Although utility for the purposes of science is the most important criterion for evaluating any activity in science, including alternative forms of theory, there are other criteria worth considering: Which of these theoretical forms is most useful for describing or explicating a new idea or paradigm, perhaps one that requires new concepts? Which of these forms of theory allows for the most efficient research? How would scientific knowledge be organized if each of these conceptions of theory were adopted?

If the problem of describing new "ideas" or new paradigms is considered, it would appear that the axiomatic-causal process form is superior to the set-of-laws form of theory. Since all concepts used in a law must be measurable in concrete settings, the use of hypothetical or unmeasurable concepts is prohibited by the set-of-laws form. But more important, it appears that new paradigms lend themselves to the causal process form of description rather than to a set of laws. These two factors, the prohibition of imaginary concepts in a law and the "natural" tendency of individuals to generate new ideas in the causal process form, would suggest that the axiomatic-causal process form of theory is more suited to the description of new paradigms.

It is possible to divide research strategies into two classes, "research-then-theory" and "theory-then-research" (described in more detail in Chap. 7). The axiomatic-causal process form of theory appears to enable more efficient research if the theory-then-research strategy is adopted. Using this strategy, a theory is described, statements are selected for empirical testing, and research is designed to determine their usefulness. If the theory is an interrelated set of statements, such as in the axiomatic or causal process form, then support for any one statement provides indirect support for the other statements in the set. However, if the statements are considered to be a set of laws, then they are independent, and support for one law does not provide support for the other statements in the set of laws. Using this strategy, the axiomatic-causal process form is more efficient in terms of research, since less research would have to be done to test a causal process or axiomatic theory than a set of laws.

On the other hand, the research-then-theory strategy in science is a procedure for conducting research and then attempting to infer what systematic patterns among the data might be considered to be laws. Given this research strategy, it is conceivable that the set-of-laws conception of theory may lead to the most efficient form of research, since many laws may be found among the large mass of data at a low cost per law. However, since in this procedure the data is collected before the theory is developed, there is no guarantee that the available data will provide a satisfactory test of the resulting theory if the axiomatic-causal process form of theory is employed. If this is the case, then a second research project would be needed to actually test the usefulness of the axiomatic-causal process form of theory.

In summary, research efficiency is related to both the conception of theory and the strategy employed for developing scientific knowledge. A conception of theory as a set of laws will lead to an efficient use of resources if the research-then-theory strategy is employed, and a theory in axiomatic or causal process form will lead to an efficient use of resources if a theory-then-research strategy is employed.

Finally, it is worthwhile to consider the way scientific knowledge is organized in the different forms of theory. If the set-of-laws conception of theory is adopted, then scientific knowledge is essentially a catalogue of theoretical statements, each one describing the relationship between two or more concepts. This catalogue of statements will have two interesting characteristics: (1) it will be *very, very* long, and (2) it will be difficult to determine the interrelationships, if any, between statements.

In contrast, the axiomatic-causal process form of theory suggests that scientific knowledge be organized around the different theories.[6] Since each theory contains a number of statements, this will undoubtedly result in a much smaller number of "things" to be organized, although the total number of statements may not be reduced. Moreover, considering theories

6. As generally presented, as a set of definitions with few propositions, "modern systems theory" (Buckley, 1967) does not result in many major predictions about phenomena and therefore is not very useful for the purposes of science. On the other hand, sophisticated analysis often suggests that sets, or systems, of causal processes exist in many situations. Sometimes these sets of processes produce "feedback systems," generally diagrammed as follows:

$$X \rightarrow Y \rightarrow Z \quad (Z \rightarrow X)$$

This can be considered as two interrelated causal processes:

$$X \leftarrow Z$$
$$X \rightarrow Y \rightarrow Z$$

With such systems the question often arises: "When will this system be stable?" or "When will there be no further change in the values of *X*, *Y*, and *Z*?"

Charles A. McClelland (in a personal communication) has suggested that general systems theory is basically a strategy for approaching social and human phenomena as a "system" of causal processes, rather than expecting *one major process* to cause everything. If this is the case, then the perspective of general systems theory is identical to the view adopted here, that the final explanation will be a set of interrelated causal processes.

as causal processes may make it easier to discover and examine the interrelationships between processes, since processes dealing with similar concepts and statements may be interrelated. For these two reasons—the small number of "things" to organize and the relative ease with which interrelationships between theories can be determined—the axiomatic-causal process form of theory is more desirable in terms of organizing scientific knowledge.

There is one final difference between the two forms of theory on the question of objectively determining when a theory is complete. Using the set-of-laws concept of theory, it is easier for a researcher, by himself, to determine whether or not he has "discovered" a law, particularly if he puts considerable emphasis on statistical decision procedures (see Chap. 6). If a scientist is attempting to develop a theory in the axiomatic-causal process form, it is more difficult for him to determine, for himself, when he has "finished"—when he has specified all the statements necessary to describe the causal mechanism. It is necessary for him to seek his colleagues' evaluations of his efforts in order to determine if he has developed a complete theory. However, since other scientists will decide whether or not to adopt a given theory as part of scientific knowledge, this can be considered the first stage of the adoption process.

In summary, the axiomatic-causal process form of theory is to be preferred over the set-of-laws model for the following reasons:

(1) It provides a sense of understanding.
(2) It makes it easier to describe new paradigms.
(3) It may allow for more efficient research.
(4) It suggests a more concise and interrelated organization of scientific knowledge.

These advantages seem to outweigh the single disadvantage of the axiomatic-causal process form—the subjective evaluation of its degree of completeness in comparison to the relatively objective evaluation of a law.

Despite these disadvantages, the set-of-laws conception of theory is by no means dead. It has many friends, reflected in the large number of propositional inventories that are being produced. The best known is by Berelson and Steiner (1964), which includes hundreds of theoretical statements, all of which have the status of laws, empirical generalizations, or hypotheses, depending on which scientist is evaluating the statements. In a more philosophical vein, Burgess (1968, p. 339) quotes Hamblin (1966) as suggesting a research strategy in which the final product is "the summarization of the obtained relationships by fitting the appropriate mathematical equations." Presumably, once one has the appropriate equations, one has scientific knowledge and can go on to study new phenomena. Those who are friends of the set-of-laws form seem to be the same individuals who have adopted a particular research strategy for the development of scientific knowledge, referred to as the Baconian concept of science and discussed in Chapter 7 as the research-then-theory strategy.

SIMULATION OR MODEL BUILDING

Scientists frequently engage in another activity, referred to as developing simulations, models, or even "representational models" (Berger *et al.,* 1962, pp. 37–66), which is often considered "theory building." Since the labels are not used consistently, they are of little help. But there is another type of activity that is different from constructing theories. This other activity is to develop a process that will reproduce the same patterns of empirical data that are found in specific concrete situations. Although this may seem to be the same as the development of a causal process theory, there is one important difference. Any process will be adopted by the simulator that will allow him to match the empirical data, even if it is obvious that there is no relationship between the simulation process and the actual causal processes that produce the empirical data. A scientist interested in developing a causal process theory would insist that each statement in the process be related to some event in the concrete setting, even if it could not be measured (as is the case with a hypothetical concept).

Simulations are often used for the solution of practical problems. Consider the task of designing a distribution system for supplying natural gas to a community, composed of a network of different-sized pipes and pumping stations. If the situation is complex, there may be several possible alternatives for the system, and it may be impossible to develop equations, or a mathematical model, to represent the characteristics of the different systems, such as the flow of the natural gas and the pressures at different points of the system. One solution may be to use another less expensive physical system, with characteristics similar to the natural gas system, to study different arrangements of pipes and pumps.

One way this is done is through the use of blotter paper and colored water. The change in the shading of colored water as it is absorbed by blotter paper approximates the change in flow and pressure of natural gas flowing through a pipe. The blotter and water system is similar to or simulates the natural gas and the supply pipes. By cutting out the proposed distribution systems in blotter paper, using different widths of paper to represent different-sized pipes, and by supplying colored water to the "supply pipe," the flow of natural gas in several different systems can be studied, and at low cost.

This is a true simulation because there is no question that the physical processes involved when water is absorbed by blotter paper are quite different from the physical processes involved when natural gas is pumped through a system of pipes under pressure. However, the relationship between the differences in water density and the differences in gas pressure are similar enough to allow the blotter paper system to *simulate* the physical system of pipes. One system is used to simulate the relationships found in another, more expensive, system.

In a similar fashion, a social scientist may invent processes to explain social situations without actually expecting the simulated processes to represent the real causal mechanism. For instance, in studying the mobility between three cities, a scientist may assume that the decision to move to a particular city is made on a probabilistic basis. The following model might be developed (Berger and Snell, 1957). Three cities, *X*, *Y*, and *Z*, are represented by the three circles. Each arrow represents movement from one city to another (or no movement, represented by an arrow that starts and stops at the same city) during a given year. The decimal numbers represent the proportion of

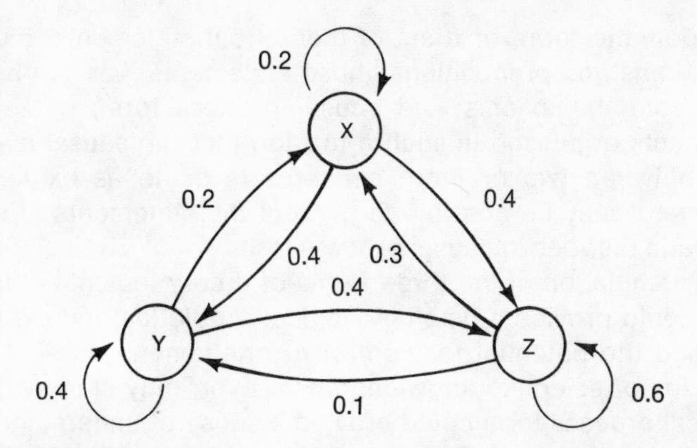

the population in each city that will move as indicated by the arrows. Whether or not a given family will move is assumed to be probabilistic.

The authors of this intercity movement model do not argue that this represents the actual causal processes in the situation. After all, given individuals may move for diverse reasons—change of job, death in the family, homesickness, and the like. However, this simulation may be used to answer a number of questions: How big will each city be when and if a stable situation is reached (i.e., when each movement from one city to another will be offset so that the sizes of the cities do not change)? How long will it take for a stable state to be reached? What will be the flows of movement between cities when a stable state is reached?

The difference between simulation processes and actual causal processes is not always clear, and it may be that some processes initially intended as simulation processes may later be considered as causal processes, or vice versa. It is best to consider the goals of the theorist when attaching labels to any theoretical activity.

SUMMARY

Three conceptions of theory have been discussed: the set-of-laws form, or the view that scientific knowledge should be a set of theoretical statements with overwhelming empirical support;

the axiomatic form, or a set of theoretical statements, divided into axioms and propositions, those statements that can be derived from the axioms; the causal process form, or sets of statements organized in such a fashion that the causal mechanism between two or more concepts is made as explicit as possible. It may be possible to present the statements of some axiomatic theories in causal process form.

Examination of the three forms of theory indicates that all three could provide a typology, logical prediction and explanation, and the potential for control of phenomena, three of the purposes of scientific knowledge. However, only statements in a causal process form could provide a sense of understanding.[7] In addition, the axiomatic and causal process form of theory seems to facilitate the explication or description of new paradigms, to allow for the more efficient use of resources in research, and to enable concise and interrelated organization of scientific knowledge. For these reasons the causal process form, or axiomatic theory in causal process form, appears to be a more useful model for a scientific theory.

In addition, process simulation, wherein the goal of the researcher is to reproduce the same empirical results found in a concrete situation, was contrasted with theory building, wherein the goal of the researcher may be to describe the actual processes that cause a phenomenon in concrete settings.

7. Dave Thomas has suggested in a private communication that those committed to model building often feel that a sense of understanding has been achieved when predictions from a formal mathematical model compare favorably with the data. In other words, a sense of understanding, or perhaps a feeling that an answer has been found, is produced when predictions from an existing logical system match the results of empirical research, regardless of the basis for the predictions.

6. Testing Theories

The most important criterion for evaluating the usefulness of any statement for the purposes of science is the degree of correspondence between the statement and the results of empirical research. Generally speaking, it is not possible to examine the correspondence between empirical research and *all* the statements that comprise a theory simultaneously. Therefore, most research projects are designed to provide evidence for the usefulness of one or a few statements. If these statements are part of an interrelated set of statements, if they are part of a theory in axiomatic or causal process form, then the empirical support for them increases confidence in the entire set of statements, the theory.

Since most research provides support for individual statements, a general discussion of how changes in confidence toward individual statements occur will be presented first, followed by a discussion of changes in confidence in theories—interrelated sets of statements.

ABSTRACT STATEMENTS AND CONCRETE RESEARCH

When properly formulated, scientific statements are abstract and thus independent of any unique spatial and temporal setting. For instance:

> The more uniform the attitudes of group members, the more cohesive the group.

On the other hand, all empirical research is conducted in specific spatial and temporal settings. Consider the following description:

> During the fall term of 1965, couples dating at *XYZ* University who had the same political orientations, both liberal or both conservative, dated more frequently than couples with unlike political orientations, one conservative and one liberal.

Differences between these types of statement raise two questions: How does concrete research provide support for the abstract statement? How can an abstract statement be proven true?

The first issue has already been discussed in Chapter 4. If it is agreed that the facts described by any concrete statement are also described by the more abstract statement, then any empirical support for the concrete statement also provides support for the abstract statement. This is also the case if it is agreed that the procedures (operational definitions) used to measure the concepts in the concrete statement are related to the meaning of the theoretical concepts. The example above demonstrates this point. Dating couples are groups, political orientation is an attitude, and frequency of group formation (dating) is a measure of group cohesion.

The second issue is slightly more complex. If a statement is abstract, it must be applicable in the future. If it is applicable to future situations, the possibility exists that it may be proven false in the future. Therefore, it is never possible to prove an abstract statement true for all possible situations as long as future situations are within the scope of the statement.

On the other hand, since abstract statements are applicable to the past and the present, it is possible to prove that they do not apply to some situations—that is, it is possible to falsify an abstract statement. This is why sophisticated discussions of research strategy often concentrate on the falsification of hypotheses rather than their verification, or proving they are true, as the goal of a research project.

But perhaps of more importance is that scientists do not consider abstract statements as either true or false. Empirical research affects the scientist's degree of confidence in the usefulness of an abstract statement rather than causing a scientist to accept or reject a statement as "true" or "false."

Two problems confuse this subtle but important issue. First, it is possible that a concrete hypothesis may be true or false, since it is related to a unique spatial and temporal setting.

The effect of concrete findings on the scientists' confidence in abstract statements will be considered in the next section. Second, the use of classical statistics, in its present form, emphasizes a "true" or "false," accept or reject, kind of decision. Although statistical decision procedures can be applied to a particular concrete hypothesis, the results are often applied directly, and inappropriately, to abstract statements. The use of classical statistical decision procedures will be discussed below (pp. 121–127).

EMPIRICAL RESEARCH AND CONFIDENCE IN ABSTRACT STATEMENTS

If empirical research can never prove an abstract statement true, then how does it affect the status of a theoretical statement? In short, the results of empirical research in concrete settings affect the confidence a scientist has in the usefulness of an abstract statement for the purposes of explanation and prediction. The aim of this section is to describe how this process operates.

Consider the following abstract statement:

> The more uniform the attitudes of group members, the more cohesive the group.

This statement will be referred to as statement *X* in the following discussion. As an abstract statement, it is impossible to prove that it is true, but it may be possible to prove that it is false.

Consider a research project designed to test the usefulness of statement *X*. The research is designed to determine if the following concrete statement, considered to be an instance of statement *X*, is an accurate description of actual behavior.

> During the fall term of 1965, couples dating at *XYZ* University who had the same political orientations, both liberal or both conservative, dated more frequently than couples with unlike political orientations, one liberal and one conservative.

Since this statement applies to a particular group of people at a particular time, it could be true or false.

As a sample situation, consider the two scientists, Jones

and Smith. Jones conducts the research related to general statement *X*, and the effects of his research on Smith's attitude toward statement *X* will be considered. The effect of Jones's research on Smith's attitude toward statement *X* will depend on at least three factors:

(1) Smith's attitude toward statement *X* before he learns the results of Jones's research.
(2) The correspondence between statement *X* and Jones's research findings.
(3) The care and sophistication of the procedure Jones used to gather his data.

If Jones has a sloppy research procedure, Smith may be skeptical of any results Jones claims for his study. For this reason, the following discussion will assume that Jones has an acceptable research procedure and that Smith accepts Jones's results as a stable finding (i.e., a result that would occur again if the same procedure were repeated).

For a first case, assume that Smith has no real opinion about statement *X*; he considers it interesting but is neutral about its usefulness as a part of scientific knowledge. The only factor remaining is the outcome of Jones's research. There are three possible outcomes for any research: either it supports the statement, it does not support the statement, or it is inconclusive.

If Jones's results are inconclusive, then it is unlikely that Smith's attitude toward statement *X* will change. Further, this is the worst possible outcome, since it means that Jones has spent considerable time and money and has failed to answer the research question. However, it may not be Jones's fault—science is an unpredictable business.

If Jones's results either support or do not support the statement (the empirical results do or do not correspond with statement *X*), then his results will have a maximum effect on Smith's attitude. Smith will change from a neutral position to one of moderate confidence that statement *X* is useful or is not useful, depending on Jones's exact results.

For comparison, consider a second situation, in which Smith knows of five studies, all carefully done, that are consistent with statement *X*. As a result, Smith has high confidence in statement *X* as a useful descripion of the phenomenon. What effect will Jones's study have on Smith's attitude under these conditions?

If Jones's study is inconclusive, then it is unlikely to have any effect on Smith's attitude toward statement *X*, although it may affect Smith's attitude toward Jones as a scientist. With so much previous research on the phenomenon, Smith may wonder why Jones couldn't get a conclusive result.

If Jones's study is consistent with statement *X*, it is also likely to have little effect on Smith's attitude toward statement *X*, since he was already highly confident that the statement was useful. Furthermore, Smith may even ask why Jones bothered to conduct the research, since "We already knew that."

However, if Jones's results do not support the statement, he should prepare himself for conflict. Smith has high confidence that statement *X* is useful, since there are five "good" studies consistent with the statement. If Jones presents results that are not consistent with statement *X*, Smith will most likely examine Jones's research procedure in great detail before he will accept Jones's conclusion that statement *X* is not useful for that situation. If Smith is satisfied that Jones's procedure is satisfactory, then his attitude toward statement *X* may change —his confidence in the usefulness of statement *X* may be reduced.

However, other changes are not only possible, but also more likely. In these kinds of situation, where new evidence is inconsistent with accepted "beliefs," the accepted beliefs, in this case statement *X*, often undergo careful reexamination. The abstract concepts, the conditions under which the statement is applicable, the operational definitions, the relationships between concepts, and any other feature of the statement and the research that supports it come under close scrutiny. Smith and others like him will be unwilling to reject a statement that is supported by five good studies but are also unable to ignore Jones's inconsistent results.

Frequently a new statement, statement *Y*, may be developed that is consistent with the five good studies as well as Jones's results. This is the type of situation where scientific knowledge becomes more precise and sophisticated.

In summary, it is never possible to prove an abstract statement true since it could be false in the future. This leads scientists (1) to attempt to prove abstract statements false, since this can be done, and (2) to classify abstract statements in terms of scientists' confidence in their usefulness. The effect of empirical research on a scientist's attitude toward abstract statements depends on (1) the results of the research, (2) the quality of the research procedure, and (3) the scientist's attitude toward the abstract statement before he learns of the new findings.

STATISTICAL DECISION PROCEDURES

Statistics or, more precisely, statistical decision procedures are often used to "prove" whether or not a statement is true or false. Although statistical procedures are useful, they are frequently misused and misapplied. An understanding of the nature of statistical decision procedures helps to prevent their misuse.

A major misuse of statistical tests is to apply them directly to abstract statements to prove whether the statement is "true" or "false." This is inappropriate for two reasons. The first is that an abstract statement cannot be proven true, since it is applicable in the future. The second is that there is only one situation in which a statistical test can be directly applied to an abstract statement—when there is one and only one operational definition for each concept in the statement. If there is more than one operational definition for a concept, then the statistical decision procedure is applied to an operational statement that is considered to be subsumed under the abstract statement (see Chap. 4). Even in this case, the statistical procedure is not applied to the statements directly, but to a concrete statement that is subsumed under the abstract statements.

For instance, a statistical decision procedure *cannot* be applied to this statement:

> The more uniform the attitudes of group members, the more cohesive the group.

Nor can it be applied to this statement:

> During the fall term of 1965, couples dating at *XYZ* University who had the same political orientations, both liberal or both conservative, dated more frequently than couples with unlike political orientations, one liberal and one conservative.

But it *can* be applied to the following statement:

> Forty couples were selected at random at a school dance, and the 25 with similar political orientations had dated each other an average of 3.5 times whereas the 15 couples with unlike political orientations had dated an average of 1.75 times. (A variance of 0.50 was associated with the average number of dates for both sets of couples.)

This is a specific concrete example of both the operational statement and the abstract statement. If there is only one way to measure the abstract concepts, then a statistical procedure may be used to show that the abstract statement is "false," i.e., it does not describe this situation. If there is more than one way to measure the abstract concepts, then all the possible measures must be taken into account before the statement can be considered "false" or not true for that particular situation.

Classical Statistical Inference[1]

The term "statistics" is often applied to two different types of activity: descriptive statistics and inferential statistics. The goal of descriptive statistical procedures is to describe some characteristic of an event or phenomenon. For example, if one desires to measure the attitudes of all the voters of a nation toward a political candidate, it is rather expensive to ask *every* voter his opinion. Using the procedures of descriptive statistics, it is possible to get an *estimate* of the attitudes of *all* the voters by measuring the attitudes of a small proportion of the voters (selected by some random procedure). Without such descriptive statistical procedures, much of current social science research (particularly survey research) would be prohibitively expensive and could not be conducted. Other data analysis procedures, such as describing the relationship between two or more variables

1. This section may be omitted without destroying the continuity of the book.

in quantitative terms (with a number), could also be included under the label of descriptive statistics—where the goal is to describe "something."

Inferential statistics, on the other hand, is a procedure for helping individuals decide which of several descriptions of an event or phenomenon is the "true" description. This procedure is variously called "statistical inference" (making inferences about the true state of affairs) or "statistical decision" (making decisions about which description to accept). The most common of these statistical decision procedures is the classical decision procedure, a procedure that requires all value judgments, or subjective judgments, to be made before the actual decision procedure is applied. Although the procedure has numerous variations, depending on the nature of the situation and the phenomenon, the logic of this procedure is always the same.

Classical statistical procedures are best viewed as procedures for making decisions about the "true" description of "nature." The procedure can be outlined as follows:

(1) Two descriptions of the phenomenon are proposed, description *X* and description *Y*.
(2) Assuming that only one description is accurate, a procedure is developed to determine which description will be adopted as *the* true description.
(3) Four possible outcomes of this procedure are possible:
    (a) Nature is really like description *X*,
        description *X* is accepted as the true description.
    (b) Nature is really like description *Y*,
        description *Y* is accepted as the true description.

    If either outcome a or b occurs, no error has been made.
    (c) Nature is really like description *X*,
        description *Y* is accepted as the true description.
    (d) Nature is really like description *Y*,
        description *X* is accepted as the true description.

    If either outcome c or d occurs, then the wrong description has been accepted as the true description, and an error has been made.
(4) For ease of discussion, these two types of mistakes are given labels.

(a) Type I error: Nature is like *X*, description *Y* is accepted.
(b) Type II error: Nature is like *Y*, description *X* is accepted.

(5) The classical decision procedure is such that it is impossible to prevent, in an absolute sense, either type of error from occurring. However, it is possible to adjust the probability that these errors will occur. For ease of discussion, these probabilities are given labels.
(a) Alpha: The probability that a Type I error will occur.
(b) Beta: The probability that a Type II error will occur.

(6) Ideally, both Alpha and Beta should be very small, i.e., the probability of making either type of mistake should be small. Unfortunately, it is the nature of the decision procedure that these two probabilities usually have an inverse relation (if Alpha is small, Beta is large; if Beta is small, Alpha is large). The only way to reduce both Alpha and Beta is to have a large sample size.

(7) Based on the assumption that the decision procedure should be completely objective, i.e., that no subjective judgment should affect the decision procedure, the scientist is requested to select the values of Alpha and Beta he finds acceptable before he applies the decision procedure. The scientist must consider what mixture of risks he finds acceptable. Does he want a mistake to cause him to "accept" description *X* or *Y* as the true description?

(8) All the information required to use the classical statistical decision procedure has been specified. There are four types of information:
(a) Description of nature *X*.
(b) Description of nature *Y*.
(c) Alpha, the probability that description *Y* will be accepted as the true description when nature is really like *X*.
(d) Beta, the probability that description *X* will be accepted as the true description when nature is really like *Y*.

(9) The research is then conducted, the decision procedure

applied to the data, and the scientist is told which description he should accept, *X* or *Y*. Presumably, he accepts that description as being *the true description,* since he specified the risks he wanted to take.

The details of the classical procedure are sophisticated and complex, since the decision to be made (accept description *X* or *Y*) for any possible pattern of the data needs to be specified in advance of the decision. Standard decision procedures are prepared in advance and allow the scientist to select a procedure appropriate for his research situation. All of these procedures have one characteristic in common: *Description* X *is a description of nature as a random process.* The assumptions about the random process differ for different statistical tests (t-test, Chi-square, normal distribution, and so on), each appropriate for a different situation. Although this procedure has the advantage that it provides the same standard of comparison for all statistical tests, it leads to some important problems.

One problem is caused by the fact that conservatism in scientific inference is considered a "good" thing and is reflected in a tendency to accept a low value of Alpha and a corresponding high value of Beta. If we assume that the scientist's description of nature, description *Y,* describes nature as having a systematic pattern, the probabilities of making a mistake, Alpha and Beta, can be described as follows:

Alpha: The probability that a description of nature as a systematic pattern (*Y*) will be accepted when it is really random (*X*).

Beta: The probability that a description of nature as random (*X*) will be accepted when it really has a systematic pattern (*Y*).

Discussion of whether or not certain results are significant at the 0.05 or 0.01 level refer to the value of Alpha. The problem is that most users of statistical tests are unaware of the risk of Beta and often unknowingly accept a high value of Beta, in excess of 0.50 for small sample sizes, in order to achieve a low value of Alpha. This favors the description of nature as a random process at the expense of the description of nature as a systematic pattern.

The major reason that most scientists consider it prudent to adopt a low value of Alpha appears to be a basic distrust of their colleagues' research procedures. It is well known that a scientist can influence the results of research, often unconsciously, so that the results tend to agree with his hypothesis—nature has the pattern he proposed.[2] In a partial attempt to correct for these influences, it is considered prudent to insist upon a low level of statistical significance, or Alpha, even at the expense of a high value of Beta.

It is also unfortunate that the classical statistical procedure results in an "accept-reject" or "true-false" kind of decision. Typically some value of Alpha, such as 0.05, is selected as the level of significance that must be achieved before a description of nature as a systematic pattern will be accepted. It is questionable whether findings statistically significant at the 0.052 level (not quite 0.05) should be "rejected" when findings at the 0.048 level are "accepted." This is one feature of the classical statistical decision procedure that is not slavishly followed. Scientists are usually willing to consider results that approach statistical significance rather than ignore them just because they are a few decimal points higher than some arbitrary level. In fact, a strong argument can be made for reporting the level at which a finding is "just" significant rather than the level reached (0.05, 0.01, 0.001, and so on), so the reader can make his own evaluation of the "significance" of the results.[3]

Perhaps the most serious problem with emphasis on sta-

2. See Rosenthal (1966) for a discussion of this problem in experimental studies, the most highly controlled of all social science research.

3. The classical statistical decision procedure, described above, requires the scientist to make explicit his research hypothesis and the level of risk, represented by Alpha and Beta, that he finds acceptable, all in quantitative form. An alternative statistical decision procedure requires the individual to explicitly describe only two items of information, the research hypothesis and his confidence that the hypothesis is true, both in quantitative form. The individual describes his confidence by giving his estimate of the probability that all possible outcomes will occur. After applying this other statistical decision procedure, which makes some assumptions about how subjective judgments are changed, the result is the amount of confidence the scientist *should have* in the hypothesis after the new research evidence is taken into account. This alternative procedure is called the Bayesian inference procedure and is one formalization of how subjective judgments are changed, reflecting the philosophy in this book. An introduction to Bayesian inference is provided by Raiffa (1968), and a more sophisticated discussion of applications to social science is presented by Edwards *et al.* (1963).

tistical tests is that evaluation of research tends to focus on statistical significance at the expense of substantive significance. Consider the following imaginary research results:

(i) A study of the effects of lighting on the typing productivity of 100 secretaries found that when lighting intensity was increased from average to high, typing errors decreased by 2% with no loss of typing speed. This result was statistically significant at the 0.01 level (value of Alpha).

(ii) A study of the effects of supervision on five secretaries indicated that when the supervisor was absent typing speed dropped by 30% with no change in typing errors. This was statistically significant at the 0.10 level (value of Alpha).[4]

Clearly, the first study has a higher level of statistical significance; lighting clearly affects typing productivity, but in a small way. But the second study has much greater substantive significance, since the absence of supervision has a major impact on typing productivity, even though the statistical significance is above the generally accepted value of 0.05. An overemphasis on statistical significance would cause one to overlook the dramatic effect of supervision on typing productivity and focus on the minor effect of lighting intensity.

Most scientists, often intuitively, seem to realize that classical statistical procedures are not useful in the direct evaluation of abstract theoretical statements. Classical statistical procedures are used as evidence of patterns in concrete situations. The empirical results are then evaluated separately with respect to their relation to the abstract statement. When properly applied, statistical tests provide additional evidence for evaluating abstract statements, but they should not make decisions for the scientist.

It is for this reason that good research design, the development of clear and intersubjective measures (operational definitions) of the abstract concepts, and asking *important research questions* usually take precedence over tidy statistics when scientists are evaluating the quality of research. The best

4. This could easily occur if the variation in the effect on the five secretaries was high; some secretaries were not affected by the supervisor's absence at all, others completely stopped productive work. In statistical jargon, this is referred to as high variance.

research design is one in which the results are so obvious that other scientists have high confidence in the results without considering the statistical significance.

The best research design is the one that does not require statistical analysis.

Although the evaluation of research without the use of statistical inference procedures is an ideal goal, it is seldom achieved, for two reasons. First, many phenomena would be too expensive to study without the use of certain statistical procedures, such as descriptive statistics in survey research or the analysis of variance in experimental research. In order to reduce the cost of performing research to a reasonable level, statistical procedures must be used in studying phenomena and statistical decision procedures are required in making inferences about which patterns in the data are worthy of attention, i.e., which are different from patterns expected from random processes.

Second, some phenomena simply cannot be studied without the use of some sophisticated statistical and quantitative procedures, particularly in the social sciences. Many social processes, e.g., those that cause changes in the characteristics of a society, cannot be controlled or influenced like the phenomena studied in laboratories. In order to study such processes, the scientist is forced to observe and measure the phenomenon in its natural setting. As such, any description of the relationship between concepts (variables) must rely heavily on statistical or quantitative procedures in order to make inferences about what is a "significant pattern" and what is a random pattern.

For these two reasons—the economies gained by using statistical analyses and the need to study some processes in complex natural settings—it is unlikely that the ideal of research without statistics will be achieved very often.

Should the Hypothesis Be Presented before the Data Are Examined?

One issue of importance is the temporal relationship between the statement of the research hypothesis and the evaluation of the empirical results. In other words, "Should one look at the data before stating the hypothesis to be evaluated by the data?" It would seem that this depends primarily on what type of hypothesis is being tested.

The classic description of a research design is Fisher's (1966) example of the study of a lady with a special taste for tea. The lady prefers milk with her tea and claims that when she drinks tea she can tell whether the tea or milk was placed in the teacup first. Fisher describes a research design to determine whether the lady can really tell, on the basis of taste alone, whether tea or milk was poured into the cup first. He suggests that cups of tea with milk should be prepared in both fashions, milk first and tea first, and then the lady should taste each cup and give her opinion as to which was placed in the cup first before the "answer" is revealed to her. In this way, she is stating her judgment before learning the results of the research (how the cups of tea and milk were actually prepared).

Notice that in this research situation, the research hypothesis is not "Is there a difference between the tea prepared in the two different fashions?" but "Can the lady tell the difference?" Can she make a distinction between cups of tea prepared in different ways? This is clearly an intuitive judgment of the most subtle kind, and, if one really wants to test the accuracy of a subjective judgment, the judgment, or prediction, should be made explicit before the "answer" is revealed. In other words, if the goal of the research is to measure the accuracy of intuitive judgments, then the prediction about the form of the data should precede the revealing of the "answer" to the predictor.

However, if the research hypothesis is openly and explicitly derived from a well-formulated theory, then there seems to be no reason to insist that the research hypothesis, and hence the theory, be stated before the data are collected and analyzed. Such a rule would actually prevent using data from past research projects to evaluate the usefulness of a hypothesis derived from a theory invented in the present. When research is designed to test the usefulness of a theory, by attempting to falsify hypotheses derived from the theory, there seems to be little chance that prior exposure to a research finding will influence an inanimate theory. It may affect the formulation and application of the theory, but that is part of the process of developing or inventing theories (described in more detail in the next chapter). Although it is always a pleasant surprise for a hypothesis derived from a theory to accurately predict an unexpected result, there is no reason why research data that test a theory must be collected *after* the theory is invented.

In summary, it is clear that if a research project is designed to test a person's intuitive judgments about a particular situation, then these "guesses" should be made open and explicit before the data are collected, or before the "guesser" knows the "answer." However, if one is attempting to evaluate the worth of a clearly formulated and explicit theory, then there seems to be no reason to hide the results from an inanimate set of ideas. Data collected at any time can be used to "test" hypotheses derived from the theory. In fact, data used for testing new theories often come from research completed before the theory was developed.

One often hears about the necessity for stating hypotheses explicitly *before* the data are collected and analyzed in social science. This would appear to be an absolute necessity only if knowledge in social science is considered to be the intuitive judgments of social scientists, the scientists' "feel" for the phenomena. However, if social science is considered to be an explicit body of knowledge developed and used by social scientists, then the argument presented above would suggest that this is not an important requirement.

CHANGING CONFIDENCE IN THEORIES

How changes in confidence in a statement affect confidence in a theory depends in part on the conception of a theory. If a theory is seen as a set of laws, then a change in the confidence in a single statement is identical to a change in the confidence in the theory. Those statements in which confidence is very high become the laws of the set of laws that are the theory.

On the other hand, if a theory is considered to be an interrelated set of statements, either in axiomatic or causal process form, support for a single statement derived from the theory provides indirect support for the entire theory. Basically, as scientists have more confidence in (1) each statement derived from a theory, and (2) a larger proportion of the statements derived from a theory, the confidence in the theory increases.

To illustrate how confidence in statements affects confidence in theories, consider the process described by Hopkins (1964) as "explaining" how a group member's centrality (in his interaction pattern) affects his rank (prestige) in the group. The process (described on p. 103) is presented at the top of Figure 7. Below the process are the statements that can be derived from the process, classified by Hopkins's estimate of their empirical support (described on p. 88). Hopkins considers the support for three statements as "good" and for four statements as "some."

However, he considers the following four statements to be without empirical support:

(1) If conformity, then rank.
(2) If conformity, then influence.
(3) If observability, then rank.
(4) If observability, then influence.

If these four statements are considered in isolation, they are without empirical support, and one might have little confidence in them as descriptions of events in nature.

However, considered as a part of an integrated theory, confidence in them is increased. Although the level of confidence may not be as great as Hopkins's "some," it is certainly above "none"—greater than it would be if the statements were not part of a theory.

This example illustrates the "two-directional" flow of confidence. Support for individual statements increases confidence in a theory which is considered to be a set of integrated statements. However, once confidence in a theory increases above the level of "none," untested or unsupported statements derived from the theory gain the confidence of scientists because they are part of a "supported" theory.

When compared to the set-of-laws conception of theory, one of the major advantages of the axiomatic and causal process conceptions of theory is that support for any one statement becomes indirect support for the remaining statements.

FIGURE 7. THE CAUSAL PROCESS BY WHICH A GROUP MEMBER'S CENTRALITY IS RELATED TO HIS RANK AND THE EMPIRICAL SUPPORT FOR STATEMENTS DERIVED FROM THE PROCESS (from Hopkins, 1964)

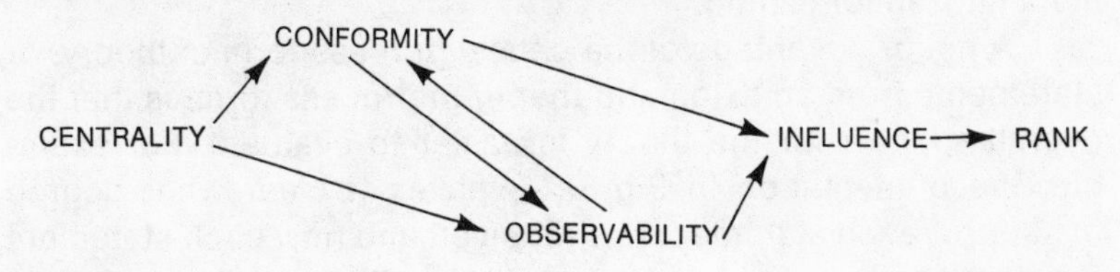

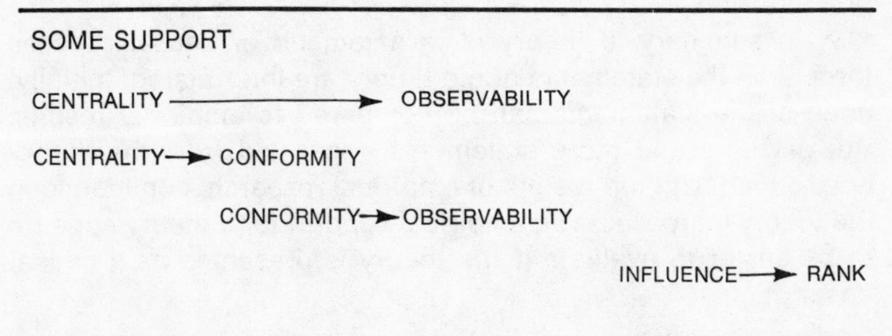

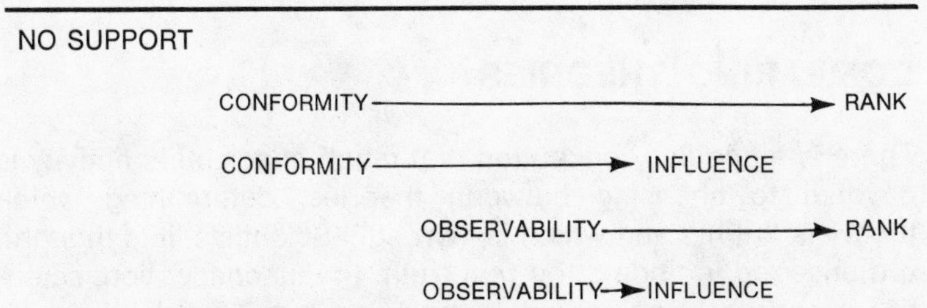

Note: All arrows represent positive correlations.

This means that research is more efficient if it tests statements from an integrated set of statements, since it affects confidence in the whole set of statements rather than in just the statement chosen for testing.

A major advantage of the causal process form of theory, or statements from an axiomatic theory in process form, is that the overall support for the theory is easier to evaluate. For example, the presentation in Figure 7 makes it clear what degree of support each statement has received and how each statement is related to the other statements in the theory. When a theory is presented in axiomatic form and the axioms cannot be tested directly, empirical support for one statement derived from the axioms provides indirect support for the axioms themselves because of the logical relation that makes it possible to derive the statements in the first place. Since the logical inference is deductive in nature—the statements are derived from axioms and propositions—the confidence in the axioms "travels backwards" or "up from" the statements. For this reason, confidence does not seem to spread so far, or so fast, among statements in an axiomatic form of theory as it does when they are in causal process form.

In summary, if theory is in axiomatic or causal process form, then the statements of the theory are interrelated. Initially, none of the statements can be compared to empirical results. But as more and more statements generated from the theory correspond with the results of empirical research, confidence in the theory increases. The empirical support for a theory appears to be easier to evaluate if the theory is presented as a causal description.

COMPARING THEORIES

There is a popular conception that much of scientific activity is devoted to choosing between theories, determining which theory is "right" and which is "wrong." Scientists, it is thought, are engaged in finding the real truth. In this conception, scientific research is perceived as composed of "crucial" experiments—experiments designed to separate the true from the false. It should be clear from the previous discussion that this conception is in error in several respects.

First, scientific activity is more usefully conceived as the development of more accurate descriptions of phenomena, not the search for real truth.

Second, no single empirical study will provide enough evidence to cause a theory to be completely rejected.

Third, if a scientific theory is thought of as a set of descriptions of causal processes, including a definition of the conditions under which they are activated, then it is more productive to ask how much effect does a particular process have under certain conditions, rather than which is *the* process (or theory) that is operating. In other words, there may be several processes that have an influence on a particular dependent variable. A more useful question is how much influence does each process have, not which is right and which is wrong.

Finally, there is another important reason why specific comparisons between theories fail to occur very often. This is because theories are usually not directly comparable. An important part of any theory or statement is a description of the conditions under which it is applicable. Only if two theories are applicable to the same situation or phenomenon can they be compared for usefulness.

However, theories or statements can be compared on two important characteristics, precision and generality. "Precision" means the accuracy of prediction whereas "generality" means the range of different situations to which a theory can be applied. Usually, these two characteristics are negatively correlated; a very precise theory is applicable to a few specific situations, and a less precise theory is applicable to a wider range of situations.

For example, consider the following two statements, either of which could be part of a theory.

(i) In any face-to-face group of interacting individuals, one individual will contribute more to the discussion than any other.

(ii) If three individuals with approximately equal societal status seek agreement on an ambiguous problem requiring subjective judgment through face-to-face discussion, one

> member will initiate approximately 45% of all acts initiated in the group (i.e., one member will do 45% of the talking). (Bales, 1970, p. 467)

It is clear that any situation that meets the conditions of statement ii will meet the conditions of statement i, but the predictions of statement ii are more precise than the predictions of statement i. On the other hand, statement i can be applied to many situations, such as larger groups, which are outside the scope of statement ii. The choice is between a general but approximate statement and a less general but more precise statement. Which is the "better" or more useful statement? The answer depends on the purpose of the scientist.

However, there are occasionally two different theories that are proposed to explain the same events, i.e., they describe the same process. If the two theories are really incompatible and are not just different processes with the same effect on the dependent variable, then it is of value to consider what criteria might be used to choose between them. Of course, it is necessary to assume that they both have equivalent empirical support. Otherwise, the theory that explains more data is the obvious choice.

One criterion frequently mentioned is that of "simplicity," selecting the most parsimonious theory. If one is using the set-of-laws concept of theory, the application of this criterion is not too difficult, particularly when the laws are being expressed in terms of mathematical equations. For instance, consider a pattern of relationships between two variables, X and Y. Three equations can be used to describe this relationship:

(i) $X = aY + b$ (linear equation)
(ii) $X = aY^2 + bY + c$ (quadratic equation)
(iii) $X = aY^b$ (power function)

If the match between the observed empirical relationships and the pattern described by these three equations is the same, then all three equations "fit" the data equally well. But the criterion of simplicity would suggest adopting the linear equation as describing the "law." Linear equations are universally considered to be simpler than quadratic equations or power functions. The choice between the quadratic equation and the power function is not so obvious.

In contrast, the situation is more complex if using an axiomatic or causal process form of theory. The major problem is that of attempting to define "simplicity" as a characteristic of an integrated set of concepts and statements. It might be possible to count the number of concepts and statements that compose the theory, but, even if there is a clear difference in the number of concepts and statements, another problem results. How do you measure the simplicity of a concept? Two "easy" concepts may be considered "simpler" than one "difficult" concept. It may be very hard to get agreement on such an issue. It is clearly not an objective decision.

A second and more important problem is that the criterion of simplicity may be inconsistent with providing a sense of understanding. If a sense of understanding is ignored as an important purpose of scientific knowledge, then simplicity might be an acceptable criterion if some procedure for measuring simplicity could be worked out. However, if a sense of understanding is achieved only when a theory provides a description of a causal process, then the most useful theory may be the one that provides the most detailed description of the causal process. In short, the most elaborate theory may provide the greatest sense of understanding. The inconsistency between the criterion of simplicity and a sense of understanding would imply that simplicity is not a useful criterion for selecting between *substantive* theories, although simplicity may continue to be useful for choosing between logical systems or the form of "laws."

If simplicity is not a useful criterion for choosing between theories, then what criterion should be adopted? It would appear that there is some reason to consider the theory that provides the greatest sense of understanding as the most desirable.

Unfortunately, a sense of understanding is almost entirely an individual matter and may vary from scientist to scientist. There seem to be at least two factors that affect a sense of understanding—the previous use of a theory (the more it has been used, the more the scientist feels comfortable with it) and the number of other theories that share the same conception of the subject matter. The second factor is particularly important with respect to the causal process form of theory. Each process can be thought of as reflecting a conception, or paradigm, of the phenomenon. If a large number of processes, or theories, share the same basic paradigm, then the sense of understanding provided by any one process is in part related to the fact that it shares the conception with the other processes. Therefore, it will be hard to reject one process, or theory, without rejecting the entire set of processes. It is for these reasons that theories do not change very fast; it often takes generations for a new theory to replace an old one.

In summary, choosing between theories is not a frequent activity in social science, but when it occurs it is one of the most important. If theories are considered as descriptions of causal processes, it may be more useful to determine the relative influence of each process, and how they affect one another, rather than determining which is *the* true process. As theories vary considerably in terms of their scope (the range of situations to which they may be applied) and precision (the accuracy of their predictions), different theories may be useful for different purposes. If a choice must be made between theories, it would appear that adopting the theory that described a causal process in the greatest detail would probably be more useful than a theory that was simpler. The intimate connection between an individual's conception of a phenomenon and the types of processes he considers as providing a sense of understanding suggests that it could take a long time for a new theory to replace an existing one.

CONCLUSION

The important points in this chapter can be listed as follows:

(1) Abstract statements cannot be proven true, but they can be proven false.
(2) Concrete statements can be proven true or false.
(3) Concrete statements provide indirect support for the usefulness of abstract statements.
(4) As abstract statements are found to be useful descrip-

tions of more and more situations, the confidence in the usefulness of the statement increases.

(5) Classical statistical tests of significance are directly applicable to concrete statements but generally cannot be directly applied to abstract statements.

(6) Substantive significance is often more important than statistical significance.

(7) Hypotheses based on intuitive judgments should be made public before research data are analyzed. Hypotheses derived from explicit theories can be tested with data collected before or after the hypotheses are stated.

(8) Support for a statement derived from a theory provides indirect support for the entire theory. As more statements are found to correspond with empirical results, confidence in the theory increases.

(9) Confidence in an untested statement is usually greater if it is derived from a well-supported theory, an interrelated set of statements, than if it is considered in isolation.

(10) It is more useful to consider "how much effect" each of several theories (causal processes) has on a phenomenon than which theory is "right," or *the* answer.

(11) Providing a sense of understanding appears to be a more useful basis for choosing between competing theories (in axiomatic or causal process form) than simplicity.

In general, confidence in theories changes slowly, with few dramatic confrontations resulting in a "winner" or a "loser."

7. Strategies for Developing a Scientific Body of Knowledge

There are and can be only two ways of searching into and discovering truth. The one flies from the senses and particulars to the most general axioms [laws], and from these principles, the truth of which it takes for settled and immoveable, proceeds to judgment and to the discovery of middle axioms [less abstract statements]. And this way is now in fashion. The other derives from the senses and particulars, rising by gradual and unbroken ascent, so that it arrives at the most general axioms [laws] last of all. This is the true way, but as yet untried.

—Francis Bacon, Aphorism XIX,
Novum Organum, 1620

A concept of scientific theory as a set of ideas useful for the purposes of science (providing typologies, explanations and predictions, potential for control, and a sense of understanding) has been developed. The procedure for describing a paradigm, or idea, using theoretical concepts and statements has been discussed. Three different ways of organizing statements to form theories were examined: set-of-laws, axiomatic, and causal process. The previous chapter described how a theory is empirically tested in terms of its usefulness as a description of natural phenomena.

This chapter will be devoted to a completely different type of discussion, a discussion of the strategies one might use in attempting to develop a scientific body of knowledge. Two basic strategies, both outlined in the statement by Bacon, have been under discussion for hundreds of years. The strategy of inventing theories that are then tested by empirical research he refers to as the "Anticipation of the Mind," and deriving the "laws of nature" from a careful examination of all the available data he calls the "Interpretation of Nature." Bacon suggests that the "true sons of science" should be using the latter method, often referred to as the Baconian strategy.

Following is a presentation of these two strategies; they will be evaluated and a composite strategy proposed.

RESEARCH-THEN-THEORY

This strategy is essentially as follows:

(1) Select a phenomenon and list all the characteristics of the phenomenon.

(2) Measure all the characteristics of the phenomenon in a variety of situations (as many as possible).

(3) Analyze the resulting data carefully to determine if there are any systematic patterns among the data "worthy" of further attention.

(4) Once significant patterns have been found in the data, formalization of these patterns as theoretical statements constitutes the laws of nature (axioms, in Bacon's terminology).

Under what conditions will this strategy, referred to as the Baconian approach, be an efficient strategy for developing a useful theory? Two conditions seem to be required. The first is a relatively small number of variables to measure during data collection. It is even more desirable if reliable measurement of these variables is both easy and inexpensive. Under such conditions, measuring "all" the characteristics of a situation or event is not a difficult task, because measuring "all" the characteristics results in a relatively small amount of data.

The second condition is that there be only a few significant patterns to be found in the data. It is then relatively easy to locate these few obvious patterns. In other words, if there are only a few causal relationships in a given situation, it may be easy to locate these relationships by examining the data.

Based on present knowledge of social phenomena, can we expect most social situations to meet these two conditions, a small number of important variables and a limited number of significant causal relationships?

NO.

If anything characterizes social science it is the lack of agreement as to what variables are important for characterizing an event or phenomenon. Hundreds of different variables are proposed and measured for investigating any particular event or entity; for example, there are easily thousands of personality variables that have been proposed for describing individuals.

This inability to get past the first step is perhaps the basic reason for being skeptical of the Baconian strategy for social science. In an absolute sense, there is an almost infinite number of ways to describe anything, and social scientists seem to have approached this with respect to many phenomena. But assuming that this problem is solved, what about the second condition, that there be a small number of causal relationships?

The second dominant feature of social phenomena is that there seems to be a relatively large number of subtle and interrelated causal relationships that influence most events. If this is the case, then it is going to be difficult to detect any systematic pattern among the masses of variables that are being analyzed in step three.

Classical statistical inference, unfortunately, does not provide an ideal means for finding such patterns in large masses of data. Assume that a statistical significance (value of Alpha) of 0.05 is selected as the level at which a relationship between two variables is considered "meaningful." This means that the probability that a scientist will infer that there is a systematic pattern in the data when the relationship is actually random is 1 in 20, if there is only *one* relationship under examination. If an investigator has hundreds of relationships under consideration, and a value for Alpha of 0.05 is adopted with a classical statistical test, then it is to be expected that 1 out of every 20 relationships will be significant at the 0.05 level, *even if there is a random relationship among the variables.* In short, when analyzing the patterns among a large number of variables, classical tests of statistical significance lose much of their value as measures of significance.

Thus the Baconian strategy appears to have two major drawbacks. First, the amount of data that can be collected is theoretically infinite, and the lack of agreement among social scientists as to what are *the* important variables has resulted in an endless list of characteristics to measure. Therefore, a scientist adopting this strategy couldn't get past the first step of listing all the important variables. Second, the problem of finding substantively interesting patterns among the resulting data is overwhelming; there are just too many potential relationships to give *all* of them serious consideration.

The question arises, if this strategy has so many disadvantages, why is it still being used? There is no question that it is still alive as a scientific strategy. A recent book on the experimental study of social processes initiates its theory chapter with a quote from Bacon (Burgess and Bushell, 1969, p. 27). The answer seems to be that this strategy is associated with two assumptions about nature and its relationship to science: (1) that there is a "real truth" to be discovered in nature, in the form of discoverable patterns or regularities, and (2) that scientific knowledge should be organized as a set of laws, reflecting the "real truth." If an individual adopts these two assumptions, it seems reasonable to conclude that research-then-"set-of-laws" is the only strategy that will result in "real" science—discovery of the true "laws of nature."

There is actually no way to test empirically the first of these two assumptions, that there is a "real truth" to be discovered in nature. The fact that confidence in some laws is so strong that they are considered to be *the* truth is irrelevant to the general proposition that truth is there to be discovered. Since it is an untestable assumption (or unfalsifiable hypothesis) and cannot be evaluated directly, it is only possible to evaluate the usefulness of strategies based on such an assumption. In a later section, the utility of the research-then-theory strategy will be discussed in more detail.

The second assumption, that scientific knowledge should be organized as a set of laws, should actually be subsumed under the first assumption. For if you assume that these laws describe relationships that are "out there" in nature, then it is reasonable to assume that scientific knowledge should be organized in such a way as to reflect the true patterns in nature. However, the relative advantages and disadvantages of the set-of-laws concept of theory have been described in detail in Chapter 5 and need not be reviewed here.

One interesting feature of this strategy is the emphasis on acquiring *all* the ideas about the phenomenon from the data. Bacon (1863, pp. 60–61) actually suggests that,

> . . . the mind itself be from the very outset not left to take its own course, but guided at every step, and the business done as if by machinery.

If the word "computer" replaces the word "machinery," this suggestion takes on a very modern tone. A number of methodological procedures have been developed for the use of social scientists that allow for the analysis of data to be done "as if by machinery"—by computer.

Factor analysis is a procedure that allows one to determine which variables "go together" in different situations. Once the procedure has indicated which measured variables are highly correlated, positively or negatively, the scientist can invent an abstract concept that will incorporate all of these operational procedures. Latent structure analysis (Lazarsfeld and Henry, 1968) uses a similar strategy but assumes a lower level of quantification of the measured variables. Measurement or scaling models (Torgenson, 1958) allow a person to collect a set of data and then, by examining the internal consistencies of the data, to determine what level of measurement (or quantification) can be attributed to a particular variable. Since many of these procedures would be impossible without modern-day computers, the "business" *must* be "done by machinery."

These mechanical techniques are not, in themselves, a reflection of the research-then-theory philosophy. But they are often used as if they will allow the "laws of nature" to be "discovered." However, a scientist may use these procedures for other purposes than implementing the Baconian strategy, science untouched by human minds; he may use them to "get ideas." More will be said about this in a following section.

As a closing comment, it appears that this strategy is not unique to scientists.

> . . . , young filmmakers, who are . . . hooked on technology, love an approach in which the thinking out in advance is minimal —an approach in which you shoot a lot of footage and then try to find your film in it. . . . The filmmaker looks for the drama in life, and may occasionally find some, but he has to pan a lot of earth for a little bit of gold. (Kael, 1970, p. 117)

THEORY-THEN-RESEARCH

This strategy may be described as follows:

(1) Develop an explicit theory in either axiomatic or process description form.
(2) Select a statement generated by the theory for comparison with the results of empirical research.
(3) Design a research project to "test" the chosen statement's correspondence with empirical research.
(4) If the statement derived from the theory does *not* correspond with the research results, make appropriate changes in the theory or the research design and continue with the research (return to step 2).
(5) If the statement from the theory corresponds with the results of the research, select further statements for testing or attempt to determine the limitations of the theory (the situations where the theory does not apply).

The major focus of this strategy is the development of an explicit theory through a continuous interaction between theory construction and empirical research.

The theory-then-research strategy is developed most explicitly by Popper in *Conjectures and Refutations* (1963), where he suggests that scientific knowledge would advance most rapidly through the development of new ideas (conjectures) and attempts to falsify them with empirical research (refutations).

If one assumes that there is no "real truth" or "laws of nature" to be discovered, but that science is a process of *inventing* descriptions of phenomena, then the "theory-then-research" approach becomes the preferred strategy. As the continuous interplay between theory construction (invention) and testing with empirical research progresses, the theory becomes more precise and complete as a description of nature and, therefore, more useful for the goals of science.

One important issue related to this strategy is the question of which statement from an axiomatic or causal process theory should be selected for comparison against empirical data. Several possibilities exist. One could (1) select the statement that is most likely to be true, i.e., correspond with the empirical results, (2) select that statement that is most likely to be false, i.e., not correspond with the empirical results, or (3) select that statement that is most crucial to the theory, i.e., most important in the formulation. If we assume that the basic purpose of scientific activity is to develop useful theories, then it would appear that the questionable or crucial statements should be tested first. Otherwise a great deal of effort may be expended on a theory that later turns out not to be useful. If the most crucial statement is tested first, or, barring that, the statement that is most likely to be false, then it will become immediately obvious where the theory needs to be changed, and, if not supported, the existing theory may be modified or a new theory may be constructed in its place. Therefore, as a general rule, research should be designed to make it as difficult as possible for theories. This doesn't mean that the research should be sloppy; it means that the weakest part of a theory should be tested first.

There is one feature of this strategy that seems to impose superhuman self-discipline upon scientists. When a theory has failed to be supported by empirical evidence, *it must be discarded or altered.* In a sense this is the hardest part of the procedure, for it takes a great deal of time and intellectual investment to develop a theory, and scientists tend to become ego-involved with *their* theories. Rejecting such a product is often very difficult, like the alchemist in the cartoon who is convinced he can get gold from lead (if the lead is "absolutely" pure). Unfortunately, at present there is no way to *force* a person to lose confidence in a theory; the human mind is always able to invent a reason why a theory "should work" or is not wrong. Even after photographs of the earth taken from the moon, there is still a Flat Earth Society. The goal of an ego-free scientific strategy may never be attainable.

There is one unsolved problem with this strategy: Where does the initial idea come from? Any theory will provide guid-

". . . and, carefully following my own infallible formula, I completed the experiment that should have resulted in gold. Such was not the case, however. I am therefore forced to conclude that the materials purchased from you three days before Michaelmas last were defective and of inferior grade."

ance for the research activity, or data gathering, and this activity will be more efficient if guided by a theory. However, much research activity might be eliminated or made less expensive if the initial theory were not just a random set of statements, an uninformed guess. In a later section (pp. 154–156) a composite approach will be discussed that attempts to combine the advantages of both of these strategies.

COMPARISON OF STRATEGIES

These two strategies, research-then-theory and theory-then-research, cannot be evaluated by any objective means because each strategy reflects a different assumption about the relationship between the "real world" and scientific knowledge. The research-then-theory approach reflects the assumption that there are "real" patterns in nature and that the task of scientists is to *discover* these patterns, the laws of nature. Like Margaret's approach in the cartoon (p. 148), there is only one way the puzzle fits together, and the only problem is to find out how to organize the pieces. Believing this, Margaret is quite content to spend a considerable length of time shuffling through the pieces because she knows that eventually she will be able to integrate all the parts in their *correct* relationship. There is *an* answer, and she will know it when she has found it.

The theory-then-research approach reflects the assumption that scientists impose their descriptions on any phenomenon that is studied. Scientific activity is the process of *inventing* theories (formalizing an idea in axiomatic or causal process form) and then testing the usefulness of the invention. Dennis is convinced that the important goal is to impose some sort of organization on the pieces, and he is content with an approximate solution, even if he must "lean" on a few pieces to make them fit, until a "better" solution is forthcoming. With his strategy, Dennis organizes the pieces in a relatively short period of time, compared to Margaret, and awaits her attempt to provide a "better" solution.

These two approaches reflect different philosophies about the relationship between nature and scientific knowledge. It

DENNIS THE MENACE

by HANK KETCHUM

Reprinted by permission of Publishers-Hall Syndicate.

may be that the final result of using the theory-then-research strategy (when no better solution can be invented) will be the same as using the research-then-theory strategy, after *all* the laws have been discovered. In this case the theory-then-research approach has the advantage of providing an approximate answer until the final truth is reached. But there is no way to develop an empirical answer to this question or even to determine if there *is* ultimately only one set of patterns or laws to be discovered. As such, these views have the same status as philosophical or religious beliefs in that accepting one strategy or the other depends on what assumptions about nature one begins with. However, the two strategies may be compared in terms of their apparent usefulness for achieving a scientific body of knowledge.

The basic problem with implementing the research-then-theory approach is that it is almost impossible to define all the variables that might be measured for any phenomenon; the list of things to measure is infinite. But even if this is overlooked, the problem of selecting the significant causal relationships from among the infinite number of possible relationships is insurmountable. There are just too many relationships to give them *all* serious consideration. However, in actual application, this strategy may not be used in this fashion. The researcher will often select only those variables to measure that he thinks might be interesting, using his professional judgment or intuition to guide his choices. Similarly, when analyzing the data, he will tend to look at a relationship between variables that he feels might be significant, again using intuition or judgment to guide his activity. However, if the scientist is using unformalized rules to make decisions about the progress of his research, it will be very difficult for another to reproduce his results, and intersubjectivity will not be achieved, thereby decreasing the confidence of others in his results.

The most fundamental problem with implementing the theory-then-research strategy is in inventing the initial theory. There is a tendency not to develop a theory in fully formalized form before the first empirical data are collected. A scientist may actually start with a brief sketch of a theory and then do some preliminary studies to determine whether or not it is worth pursuing the idea. During the preliminary research, he may actually sift through his data for any interesting patterns that have appeared that might have a bearing on the theory or the phenomenon. Again, the procedure may not be adhered to in strict form, but the focus is on theory before research—look before you leap.

The two strategies may be considered in relation to the efficiency with which empirical research is conducted. However, this depends on which conception of a scientific theory is adopted. If scientific knowledge is to be organized as a set of laws, then it is clear that the research-then-theory strategy will lead to more efficient research, since a large number of laws or empirical generalizations may be discovered from the same set of data. The more laws that can be discovered, the lower the cost per law. In contrast, the theory-then-research strategy would be less efficient for discovering a set of laws. This strategy would suggest that each potential law be stated and then a research project designed to test that single law. A separate research project for each law would be more expensive than testing many laws at once.

On the other hand, if scientific theory is to be organized in axiomatic or causal process form, it would be inefficient to test a number of propositions not relevant to the theory while testing propositions relevant to the theory. A thorough test of one or a few propositions derived from the theory may provide all the information necessary to change or reject the theory. In addition, as mentioned earlier, since the theory is an interrelated set of statements, confidence in all the statements in the set is affected by a test of any one statement. Rather than completing one large project designed to test all the statements in the theory at once, it is better to have a number of smaller research projects, and, as each project is completed, the theory can be revised and improved and then retested in the next project. Thus, which strategy is more economical depends on the conception of how scientific knowledge should be organized.

Finally, it is possible to ask which of these strategies has been used by those individuals who have been most responsible for making great advances in the sciences, natural and social. It would appear that no "scientific revolution," or Kuhn paradigm, has been developed by a scientist using the research-then-theory or Baconian strategy in his research. This is perhaps the most damning evidence against this approach. Although this does not mean that the research-then-theory strategy will not lead to scientific revolutions in the future, it does suggest that it should be deemphasized, at least in the form Bacon suggested.

In summary, if one considers that scientific knowledge should be organized as a set of laws that reflects the patterns in nature that "really exist," then the research-then-theory strategy provides for more efficient research, but (1) it will not provide an approximate answer until the final solution is achieved, and (2) it has the disadvantages of a set-of-laws form of theory. On the other hand, if one assumes that scientists invent descriptions of the real world and then test the usefulness of their inventions, the theory-then-research strategy provides for (1) more efficient research if the axiomatic or causal process form of theory is employed, and (2) for some kind of organized scientific knowledge, which is gradually improved and becomes more useful as research continues.

A question then arises. If the major problem with the theory-then-research approach is how one gets the original idea, and no major paradigm was developed by a scientist using the research-then-theory strategy, just how did scientific revolutions occur? The next section is devoted to an examination of this issue. It is followed by a description of a third strategy, a composite approach which attempts to combine the most useful elements of these two strategies.

HOW TO GET A NEW IDEA

What are the characteristics of a "new idea"? There seem to be two things that are involved:

(1) The invention of a new theoretical concept that can be used as part of a theory, either hypothetical (imaginary) or one that has empirical referents (can be measured with operational definitions). In the example of the jig-saw puzzle, this is tantamount to cutting the picture into a new set of pieces.

(2) A suggestion of new ways of organizing the causal relationships among "old" or "old" and "new" theoretical concepts. Following the analogy, this is like putting the pieces of the puzzle together in a new way.

A "new idea" or paradigm may incorporate either or both of these characteristics, but at least one seems to be present in most activities considered "creative."

This section will attempt to answer the question, "How to

get a new idea?" by examining the conditions under which new ideas seem to have occurred in science. These conditions are developed from a reading of philosophy of science, particularly Kuhn (1962), and are proposed as a tentative list, subject to revision.

First, and most important, one person is usually responsible for a new idea, particularly if it involves the "invention" of a new concept. Considering the necessity for explicitly communicating ideas in any group endeavor, there is reason to believe that this prevents the development of new and uncommunicable concepts. Any really new concept will not be easy to describe in terms of existing vocabularies, and the requirement for discussing ideas as they develop may inhibit the development of new and uncommunicable concepts (Reynolds, 1968). Although it is clear that science requires large numbers of dedicated individuals to advance, most major paradigms, Kuhn or otherwise, have been attributed to single individuals working alone.

Second, the individuals who are responsible for developing the new idea are usually pretty bright, above average. However, they are not superbrains as often depicted, even in the scientific literature. One only needs casual contact with living individuals with a "genius" image to realize that they are not human computers. However, they do have other important characteristics.

Third, these bright, solitary individuals have a clear and secure understanding of what a good idea is. They understand, perhaps intuitively, the difference between a "good" and a "bad" idea. This helps them to guide their own activities and reduces the amount of time they spend on useless endeavors. In short, they know when they have invented a potential good idea.

Fourth, they have a thorough knowledge of the existing major ideas and theories related to the phenomenon they are studying. Although this may not be a knowledge of the complete history and details of all research that has preceded their efforts, they know the currently accepted ideas and expectations about the phenomenon. If this were not true, they would have no way of knowing when they had discovered or invented a *new* idea. In short, they know when a good idea is a new idea.

Fifth, they are not slavishly committed to any of the existing paradigms, theories, or ideas currently accepted by the scientific community as *the* explanation of the phenomenon. Kuhn (1962) makes the point that those individuals that are responsible for creating new paradigms in the physical sciences are either young individuals or older individuals new to the field. This suggests a lack of commitment to existing views.

Sixth, they are very close to the phenomenon, in one way or another. This means that they are either trying to apply existing paradigms or theories to a particular body of data, attempting to elaborate some aspect of existing theories through empirical research, or attempting to integrate or improve some of the theoretical ideas developed by others. But no matter what the endeavor, the individuals are deeply engrossed in the subject matter, so deep that an intuitive or uncommunicable organization of new concepts and their relationships may develop —a "feeling" that later takes form as a theory.

The overall image is that of one intelligent individual who knows what useful ideas are, who is well acquainted with the existing theories in the area he is studying, who is not committed to any of the existing theories, and who is working closely with both the theories and the phenomenon. Perhaps as these individuals attempt to organize, integrate, and explain empirical data with existing theories, constantly evaluating their own confidence in these theories, they find themselves dissatisfied with the "fit" between existing theory and data and develop new ways of perceiving and explaining the phenomenon. The result is a new idea that the individual must translate into existing scientific language and sell to his colleagues.

This stereotype of the situation in which new ideas are developed may not be completely accurate, but it clearly differs from two competing stereotypes, suggested by the two scientific strategies discussed above. The research-then-theory strategy presents the image of an individual collecting and analyzing data like a giant computer and slowly "discovering" patterns in the data which eventually become scientific "laws." The theory-then-research strategy presents the image of an individual dreaming away in a study and finally developing an explicit and well-formed theory which he then sets out to test and refine. Neither of these is an accurate portrayal of how new ideas, or scientific creativity, actually occur.

The fact that neither of the two research strategies considered in isolation seems to represent the process of developing new ideas in science suggests that some combination of the two procedures is desirable.

COMPOSITE APPROACH

The research-then-theory strategy has the disadvantage that considerable effort may be spent on collecting data that have no useful purpose, but it may provide some information useful for inventing theories. The theory-then-research strategy has the disadvantage that the scientist may have no initial information on which to base the first attempts at a theory, but research is more efficient when one only collects information related to a few important hypotheses. A composite of these two strategies may provide a more efficient overall procedure and simultaneously provide a more accurate representation of the process that actually occurs in building scientific knowledge.

The composite approach divides scientific activity into three stages:

(1) *Exploratory.* Research is designed to allow an investigator to just look around with respect to some phenomenon. The researcher should endeavor to develop suggestive ideas, and the research should be as flexible as possible. If possible, the research should be conducted in such a way as to provide guidance for procedures to be employed in research activity during stage two.

(2) *Descriptive.* The goal at this stage is to develop careful descriptions of patterns that were suspected in the exploratory research. The purpose may be seen as one of developing intersubjective descriptions, i.e., empirical generalizations. Once an empirical generalization is developed, it is then considered worth explaining, i.e., the development of a theory.

(3) *Explanatory.* The goal at this stage is to develop explicit theory that can be used to explain the empirical gen-

eralizations that evolve from the second stage. This is a continuous cycle of:
(a) Theory construction;
(b) Theory testing, attempts to falsify with empirical research;
(c) Theory reformulation, back to step 3a.

This procedure differs from the research-then-theory approach in that data collection in the exploratory stages is not thought of as the final answer but is acknowledged as exploratory research utilizing a flexible research design. It is a procedure in which hunches and insights are expected to affect the data collection. This strategy differs from a rigid theory-then-research approach in that it assumes that a useful theory is hard to invent without some acquaintance with the phenomenon, which can be gained during exploratory and descriptive research. In a sense this acknowledges that theory construction is a difficult and time-consuming activity and should not be attempted in a vacuum.

This procedure seems to have the advantages of both of the previous procedures. If the actual research activity is tailored to the stage of the research, then rigorous and elaborate measures will not be employed during exploratory stages to collect a great deal of useless data. When a worthwhile hypothesis is derived for testing, *it* can be given a thorough examination, during theory construction. Finally, it makes allowances for the difficult and subtle activity that is involved in creating new ideas or new paradigms. By moving back and forth through these stages a person attempting to develop a new paradigm may well be able to match different stages of his thinking with the appropriate stage of the research process.

Some of the mechanical procedures described above in the research-then-theory section—such as factor analysis, latent structure analysis, and the application of scaling and measurement models—fit into the exploratory stages of the composite approach. They become useful devices for analyzing great volumes of data and determining if there are any patterns that deserve further attention during the descriptive and, perhaps, theory-construction stages of the strategy.

In summary, this composite strategy, with its exploratory,

descriptive, and theory-construction stages, appears to combine the advantages of the research-then-theory and the theory-then-research strategies. Initial research is conducted in an attempt to provide suggestive patterns that may be established by descriptive research. Once an empirical generalization is established, a theory may be constructed to explain this regularity. Resources are not wasted in gathering a lot of information expecting to find laws by searching through the data. Theories are not invented until there is some information about the phenomenon that will help in the development of a useful initial theory. Finally, when a theory is ready to be tested, a backlog of experience in doing research on the phenomenon allows for a sophisticated comparison of the theory with the empirical world.

RESEARCH METHODS

There are three primary types of research activity currently associated with social science:

(1) *Individual observation.* The researcher directly observes a certain social phenomenon in a natural setting and attempts to provide an accurate and unbiased record of his observations.

(2) *Survey.* A collection of people or social systems is measured with respect to certain individual characteristics (e.g., age, income, occupation, attitudes toward everything, and the like). If the entire collection cannot be measured, then a smaller group is selected as representative of the larger group.

(3) *Experimental.* The phenomenon (an individual or social process) is reproduced in a controlled situation, and then various measurements are made of the phenomenon, often measurements that could not be collected in natural settings.

The reason for listing these types of procedures is to emphasize that there is no logical reason for a direct correspondence between the three stages of research (exploratory, descriptive, and explanatory) and the three types of research activities (observation, survey, and experiment). Although there is a rough correlation between the two typologies, partially related to the interests of the researchers, it is in no sense a necessary correlation.

For example, many studies in laboratory, or highly controlled, settings have been exploratory in nature. A controlled setting is used (1) to allow certain types of measurements to be made that could not have been made in a natural setting, and (2) to isolate certain processes or phenomena that are confounded with other processes in natural settings. This was Bales's (1951) strategy in his early research on interaction in face-to-face groups. He used a highly controlled laboratory setting so that he could get an accurate record of who initiated and who received verbal comments in a face-to-face group discussion, an impossible task in natural situations. Although experimental research is frequently conducted in controlled settings to answer specific research questions or to test specific hypotheses, there is no justification for assuming that this is the *only* use of laboratory settings.

In a similar fashion, survey research can be used at any stage of the process—as a source of ideas for theories, as a description of the phenomenon, or as a test of theoretically derived statements.[1] However, it seems accurate to suggest that the most common use of survey procedures is to describe sets of individuals or characteristics of systems of individuals. This is true particularly with respect to practical research, designed to answer practical rather than scientific questions, such as market research, surveys of political attitudes, etc.

Direct observation appears to be the most appropriate for exploratory research, but there may be some phenomenon that cannot be studied with either survey or experimental procedures. This is particularly true when the phenomenon is so unique that there are not enough different cases to permit the use of survey techniques and the phenomenon cannot be controlled by the experimenter. Certain social processes considered too important to allow scientists to study them directly —such as the relationship between a presidential candidate and his campaign workers or the decision processes among Supreme Court justices—may also fall into this category.

What is important in evaluating any research activity—

1. The techniques of "path analysis" allow the testing of causal process models with survey data (Borgatta and Bohrnstedt, 1969, pp. 3–133).

exploratory, descriptive, or explanatory—is its relationship to the process of building scientific knowledge. No research procedure is inherently "bad" or "unscientific"; what may be bad or unscientific is the way the research is conducted or utilized in theory construction. It is regrettable that many social scientists are classified or evaluated with respect to the type of research activity they pursue, which may be irrelevant to their ability to make a contribution to scientific knowledge.

CONCLUSION

In this chapter several important points have been made. First is the suggestion that it may be more efficient, and it is clearly more like past scientific activity, to assume that theories are invented descriptions of nature rather than the "real truth" that is "discovered." Second, although it is difficult to determine the processes that cause the development of new ideas, it is clear that they are not developed by those using a Baconian, or the research-then-theory, strategy. Third, a research activity should be evaluated with respect to its appropriateness to a stage of the composite strategy (exploratory, descriptive, or explanatory).

The suggestions and discussions in this chapter are appropriate only as long as the goal of the research is the development of scientific knowledge. Frequently, research is conducted as an aid in achieving some immediate practical goal, e.g., the solution to a social problem, or in achieving some immediate effect on social phenomena, e.g., attitude research designed to improve an advertising or political campaign. For such projects the approach suggested in this chapter would be inappropriate and a variation of the research-then-theory strategy may be the most economical.

8. Conclusion

The philosophy reflected in this book is based on several fundamental assumptions.

(1) Scientific knowledge cannot explain why "things" exist, an unanswerable question, but rather why one event is related to another—how one "thing" affects another "thing."

(2) Scientific knowledge is that set of ideas that the community of scientists adopts as useful for the purposes of science—those ideas that provide:
   (a) typologies or systems of classification;
   (b) logical explanations and predictions;
   (c) a sense of understanding;
   (d) the potential for control of phenomena.

(3) Although it is impossible to prove that any substantive statement (referring to a phenomenon in "nature") is absolutely *true,* it is possible for scientists to have a high degree of confidence that a given statement is useful for the purposes of science, including predicting events in the future. Such confidence usually increases if the abstract ideas are congruent with the results of concrete empirical research.

(4) In its final form, scientific knowledge will probably be a set of interrelated causal processes, rather than one overall grand theory that explains everything.

(5) New paradigms, which provide the basis for new causal processes, provide the major advances in science, although it often takes considerable effort by intelligent and dedicated individuals to provide the details and precise descriptions that make such paradigms both empirically testable and scientifically useful.

From these assumptions, the following general suggestions with respect to scientific activity have been developed:

(1) Any ideas proposed as potential scientific knowledge, i.e., presented for acceptance by the community of scientists, should be as clear and as explicit as possible, even though in abstract form.
   (a) All concepts should be clearly defined and, if possible, the level of quantification made explicit.
   (b) The relationship between concepts in statements should be made *explicit,* and particular attention should be paid to the direction of causal relationships, whether deterministic or probabilistic, to differentiate them from statements of association.

   In short, the ideal description of any scientific idea makes it easy to know when it is wrong. Clearly presented ideas are the easiest to falsify.

(2) Theories, or sets of statements, should be developed in the form of causal processes, which achieve all the purposes of science, rather than the axiomatic or set-of-laws form, which fails to provide a sense of understanding.

(3) The development of scientific knowledge will probably be facilitated by a process of trial and error: inventing theories and revising or rejecting those that fail to be supported by rigorous empirical tests. The resulting theories become progressively more useful for the purposes of science.

The remainder of this chapter will focus on some problems that affect the potential for developing scientific knowledge related to social and human phenomena.

POTENTIAL FOR A *SOCIAL* SCIENCE

In considering whether or not a scientific body of knowledge related to social and human phenomena can be developed, two types of problems are worthy of discussion: (1) the special characteristics of social phenomena, and (2) the special characteristics of social scientists.

First, if scientific theory is considered to be a set of causal processes, then a scientific body of knowledge related to social and human phenomena will probably involve more interrelated processes than will a science of physical phenomena. Even the simplest social and human phenomenon involves a host of subtle and interrelated processes, more like a biological system than a physical system. Despite the complexity, the procedure is clear—identify and describe the major causal processes and then determine their interrelations.

Second, and also of major importance, is the problem of measurement. Many of the most useful and highly developed theories of social and human phenomena cannot be used to predict or explain events (and cannot be tested empirically) because there is no way to identify their occurrence in concrete situations—they cannot be measured. The strategy reflected in this book is to invent the abstract concepts and then to try to develop ways of measuring an instance of the concept in concrete situations. Another strategy, a variation of the research-then-theory approach, is to determine what the fundamental dimensions are, what are the "things" that can be measured, and then to build theories on these measurable concepts (see Torgenson, 1958). At present, neither strategy seems to have been a sure winner, and the problem remains—how to achieve intersubjective measures of useful theoretical concepts.

Third, social and human phenomena fight back—and in two ways. First, observing and measuring the phenomenon may actually affect the phenomenon itself, easily demonstrated by pointing a camera at anyone (if they recognize the camera). This effect is well known in both physical and social sciences, and the major problem is to determine how much this affects the phenomenon so it can be discounted as an influence on the events of interest. One solution in social research is to attempt to measure the phenomenon unobtrusively—so that the individuals are unaware that they are being observed (Webb *et al.*, 1966).

The phenomenon also fights back by arguing with the social scientists about what "really happened," even if the research investigates processes that are partly unconscious, involve physiological variables, or are social processes that are difficult for any one member of a social system to observe. If a social scientist presents his analysis and explanation and it is incongruent with the "popular" explanation, then he will probably meet resistance and often scorn. This is not so frequent in other sciences; few individuals argue with a physician's explanation of why their side hurts, and the atoms do not talk back to the physicists. Perhaps, as social science becomes more comprehensive and precise, this will occur less frequently.

A fourth problem with developing a scientific body of knowledge related to social and human phenomena is the problem of maintaining an objective or "value-free" orientation. Ideally, decisions affecting the development of scientific knowledge will reflect only a desire to achieve the goal of science—understanding. However, an individual's decisions and perceptions, even when he is a scientist trying to be honest and objective, are often unconsciously affected by a variety of subtle, systematic, and unconscious processes. For instance, it is difficult to study the processes related to stereotyping and prejudice without acquiring a bias about what is "good" or "bad." Although the ideal of a *completely* objective or value-free social science may be unobtainable, it is not impossible to expect social scientists to be open and explicit about their biases, so that others may consider them in evaluating their work.

A final problem, shared with some biological and medical sciences, is one of ethics in conducting research on humans, a problem that does not occur with research on physical phenomena. When research is conducted on humans, there are ethical restraints with respect to the risks or measurements to which they should be subjected. In experimental research, where individual characteristics may be deliberately changed to examine their influence on other variables, the possibility of any damage, temporary or permanent, must be kept to a minimum; the research subjects should be completely aware of any risks; and subjects should have the option of refusing to participate. In studies in natural settings, such as survey research or participant observation, the respondents or participants should be guaranteed privacy and confidentiality in any matter *they* would consider embarrassing. It is reasonable to conclude that some phenomena will be difficult or impossible for social scientists to study because of these ethical problems. Fortunately, those phenomena that raise ethical problems are a small percentage of all social and human phenomena.

In summary, there are five problems, inherent in dealing with social and human phenomena, that increase the difficulty of developing an empirically based body of scientific knowledge: (1) the large number of subtle and interrelated processes, (2) the problems of achieving intersubjective measurement of abstract concepts, (3) the change in many social and individual phenomena under observation and the tendency for lay individuals to resist the interpretations and explanations of social scientists, (4) the difficulty of achieving complete objectivity in dealing with social phenomena, particularly when related to sensitive issues, and (5) ethical considerations that prevent the use of certain types of research procedures or require more expensive alternatives. Many of these problems are not present in the study of other types of phenomena, such as physical or chemical processes, and some are shared with the study of other phenomena, such as human physiological processes, but this set of problems seems to be unique to the study of social and human phenomena.

However, despite these many problems, the major factor that thwarts the development of a scientific body of knowledge of social and human phenomena is the character of social scientists themselves—problems within the social scientists, not within the phenomena. Two major deficiencies are lack of clarity in theoretical writings and ignorance about what scientific knowledge should look like and how it is created.

If one is even briefly familiar with the writings of social scientists (in anthropology, political science, psychology, and sociology), it is clear that much of what is written fails to meet the most fundamental criterion for a scientific body of knowledge—it is ambiguous. In fact, many social science writings have all the clarity and precision of an astrology prediction or political speech; it is difficult to determine what processes are being described, the relative influence of different processes or factors is vague, and the nature of the predictions is often "puncture proof and self-sealing"[1] (they cover all the holes). In short, much of social science theory is so ambiguous that it cannot provide logical predictions or explanations, cannot be tested empirically, and can never be proven wrong.

This factor, in addition to other features of these writings,

1. A phrase borrowed from Robert Nisbet (1968, p. 989).

suggests that many social scientists have no clear understanding of what scientific knowledge should look like or how they might help develop such a body of knowledge. This book is an attempt, at an introductory level, to explore some of these issues and suggest useful guidelines for the development of a scientific body of knowledge related to social and human phenomena.

Given these two sets of problems, the difficulties of doing scientific work with the phenomena and the deficiencies of social scientists, is there any hope for a social science? There are two ways to respond to this question. The first is to answer it directly. In the author's opinion the answer is *yes.* Not only is there a growing body of literature that meets all the criteria of scientific knowledge, but much of existing work in social science can be placed in "scientific form," although it often takes an inordinate amount of work, usually devoted to determining what the writer meant. These two facts suggest that the problems inherent in the phenomena themselves are not insurmountable and that some individuals interested in social and human phenomena are able to treat them in a scientific fashion.

The second way to respond to the question, "Is social science possible?" is to consider the alternative. The alternative is to continue to rely on folk wisdom, judgment, common sense, etc., for explanations of social and human phenomena and as a basis for making decisions affecting humans, individually and in social systems. The inadequacy of common sense as an explanation of phenomena should be apparent by now (it is not intersubjective), and its failure as a foundation for decision making has been demonstrated repeatedly. We can do better with a social science—it should be encouraged.[2]

2. The issue of how social science can be effectively used in governing a society has only recently come under discussion, as only recently social scientists have been taken seriously. It is not yet clear what the most effective procedure, in terms of benefit to society, for utilizing their expertise will be. It is clear that they have helped as "technicians" to implement certain programs or as "troubleshooters" to help solve or suggest solutions for predefined "social problems." But how social scientists can contribute effectively at a more fundamental level, helping to determine national policies and, hence, defining the programs to be implemented or the social problems to be solved, has yet to be determined. A recent book sponsored jointly by the National Academy of Sciences and the Social Science Research Council (1969) reviews the state of the social and behavior sciences and discusses how they can be used for the betterment of society.

In conclusion, it is clear that the goal of developing a scientific body of knowledge related to social and human phenomena is complicated by a number of problems inherent in the phenomena and that social scientists are not yet "ideal" scientists. (But it is not clear that they are vaguer or more biased than physical or biological scientists.) However, considering the dramatic advances that have been made in developing a social science, even if the form is often wrong, the increasing scientific awareness of social scientists, and the unacceptability of the alternative to developing a social science, there is reason to think that scientific knowledge related to social and human phenomena is both possible and desirable.

Appendix: Student Exercises

COMMENTS

Several important activities in the construction, testing, and application of theories are essentially matters of judgment, particularly evaluating the relationship between abstract theoretical concepts and operational definitions (designed to identify or measure the concept in concrete settings) and assessing the relationship between causal process theories and empirical generalizations and hypotheses.

For these reasons, the following assignments were designed to allow undergraduates to work with these types of judgmental problems, each assignment building upon the previous one. The first assignment requires the student to describe the relationship between empirical research and an abstract empirical generalization. Later assignments require an empirical generalization or an hypothesis to be related to a theory and a theory used to explain natural phenomena. By restricting the number of pages, the student is forced to emphasize the organization and integration of the essential ideas, rather than elaborate on the trivial (often getting lost in the process). The short length actually makes the assignments more difficult.

The writer has used these assignments successfully in a substantive course (introductory social psychology), requiring the student to cover different areas of the subject matter on each assignment. The assignments are usually given two weeks apart and are promptly returned to the students; by the third assignment the students produce work of high caliber. Students seem to prefer a system whereby later papers are given more weight in determining the final grade than the earlier papers. As some of the evaluations of the papers are subjective, this author has used two graders; each grades all papers independently, and then they agree on a final grade for each paper. Giving the students access to anonymous copies of good and poor papers after the assignments are graded seems to help convey some of these subjective issues to the students. Although this requires each student to turn in three copies of each assignment, they do not seem to object—they are short assignments and the reason for three copies is clear to the students.

ASSIGNMENT I: EMPIRICAL GENERALIZATION AND EMPIRICAL SUPPORT

The purpose of this assignment is to provide experience in evaluating the support empirical research gives to an empirical generalization. You are asked to describe an empirical generalization, one type of abstract statement, the results of at least one empirical study (preferably two), and evaluate the support the studies provide for the empirical generalization.

You should have read the first four chapters of this book before attempting this assignment. The paper should be organized into the following sections:

(1) Presentation of the empirical generalization.
(2) Description of the research study or studies.
(3) Description of how the research study or studies provide support for the empirical generalization. This should include a discussion of the appropriateness of operational definitions for measuring the abstract concepts in the empirical generalizations.
(4) Evaluation of the empirical generalization, its empirical support, and your confidence in it as a useful description of a natural phenomenon. If you are not confident in the empirical generalization as stated, how would you change the empirical generalization or the conditions under which it is applicable to increase your confidence in it as a useful statement?

Complete this assignment in no more than two double-spaced, typewritten pages.

ASSIGNMENT II: EXPLANATION OF AN EMPIRICAL GENERALIZATION

The purpose of this assignment is to provide experience in working with a causal process form of theory. Such theories describe the causal process that links changes in the independent variable, or concept, with changes in the dependent variable, or concept. The paper should describe one empirical generalization, one empirical study that supports the empirical generalization, and one causal process that explains the patterns described by the empirical generalization. This can be done by describing the causal process in abstract terms and then in concrete terms, specific to the concrete setting in which the research was conducted.

You should have read through Chapter 5 before attempting this assignment. The paper should be organized as follows:

(1) Describe the empirical generalization.
(2) Describe, in abstract terms, the causal process form of theory that will be used to explain the empirical generalization. Remember to make clear any concepts that might not be understood by the reader.
(3) Describe the results of one empirical study that are consistent with the empirical generalization.
(4) Describe how the causal process, previously described in abstract terms, would occur in the research setting in concrete terms.
(5) Evaluate your confidence in the theory as a useful description of natural phenomena. Be sure to discuss the reasons for your level of confidence (high or low) and any changes you might make to increase your confidence in the usefulness of the theory.

Complete this assignment in no more than two double-spaced, typewritten pages.

ASSIGNMENT III: TESTING A THEORY

The purpose of this assignment is to demonstrate how the usefulness or validity of a theory is determined. You are asked to describe one axiomatic or causal process theory, derive one hypothesis from the theory that is suitable for an empirical test, and then design a research program that could falsify this hypothesis.

You should be familiar with this book through Chapter 6. The paper should be organized as follows:

(1) Describe any axiomatic or causal process theory, making sure that all concepts will be understood.

(2) Select one statement derived from the theory as a hypothesis for empirical test; be sure to show how the hypothesis was derived.

(3) Describe a research procedure in which the hypothesis can be falsified. Be sure to show how the abstract hypothesis is related to the concrete events, i.e., the relationship between abstract concepts and operational definitions.

(4) Suggest your next move in the event that (a) the results of the research are consistent with the hypothesis (and the theory) and (b) the results of the research are *not* consistent with the hypothesis (and the theory). Assume that your goal is to develop a theory in which you can have confidence.

Complete this assignment in no more than two double-spaced, typewritten pages. Do *not* discuss any specific statistical procedures unless instructed otherwise. You may wish to read Donald T. Campbell, "Factors Relevant to the Validity of Experiments in Social Settings," *Psychological Bulletin* (1957), *54:*297–312 (Bobbs-Merrill Reprint S–352), or Hubert M. Blalock, Jr., *An Introduction to Social Research* (Englewood Cliffs, N.J.: Prentice-Hall, 1970; 117 pages) in preparation for this assignment.

ASSIGNMENT IV: APPLICATION OF THEORIES TO NATURAL PHENOMENA

The purpose of this assignment is to demonstrate how a theory is used to explain or predict phenomena. Your assignment is to select one natural phenomenon, some event in the real world, and describe two theories that will explain this event. Remember theories cannot explain why anything exists, only how one concept, or variable, is related to another concept, or variable.

You should have read through Chapter 6 of the book before attempting this assignment. The paper should be organized as follows:

(1) Describe the "natural event" you wish to explain in concrete terms.
(2) Describe the two theories you will use to explain this event; make sure that all concepts in each theory are clearly described.
(3) Show how the abstract theories explain or are related to the natural event. Be sure to describe how the abstract theoretical concepts can be identified in the concrete situation.
(4) Evaluate your confidence in these two theories as (a) useful for explaining this phenomenon, and (b) useful as *the* complete explanation of this phenomenon. Are there other theories that might also apply to this situation?

Complete this assignment in no more than two double-spaced, typewritten pages.

References

Anderson, Theodore R., and Morris Zelditch, Jr. *A Basic Course in Statistics: With Sociological Applications.* 2nd ed. New York: Holt, Rinehart & Winston, 1968.

Bacon, Francis. *The Works of Francis Bacon: Novum Organum.* Vol. VIII. Translated by James Spidding *et al.* Cambridge, England: Riverside Press, 1863.

Bales, Robert Freed. *Interaction Process Analysis.* Reading, Mass.: Addison-Wesley, 1951.

———. *Personality and Interpersonal Behavior.* New York: Holt, Rinehart & Winston, 1970.

———, and Philip E. Slater. "Role Differentiation in Small Decision-Making Groups," in *Family, Socialization, and Interaction Processes.* Edited by Talcott Parsons *et al.* Glencoe, Ill.: The Free Press, 1955.

Barber, Bernard. "The Resistance of Scientists to Scientific Discovery," *Science* (1961), *134*:596–602.

Berelson, Bernard, and Gary A. Steiner. *Human Behavior: An Inventory of Scientific Findings.* New York: Harcourt, Brace & World, 1964.

Berger, Joseph, and J. L. Snell. "On the Concept of Equal Exchange," *Behavioral Science* (1957), *2*:111–118.

———, Bernard P. Cohen, J. Laurie Snell, and Morris Zelditch, Jr. *Types of Formalization in Small-Group Research.* Boston: Houghton Mifflin, 1962.

Blalock, Hubert M., Jr. *Theory Construction: From Verbal to Mathematical Formulations.* Englewood Cliffs, N.J.: Prentice-Hall, 1969.

———. *An Introduction to Social Research.* Englewood Cliffs, N.J.: Prentice-Hall, 1970.

Blum, Gerald S. *Psychodynamics: The Science of Unconscious Mental Forces.* Belmont, Calif.: Brooks/Cole, 1966.

Borgatta, Edgar F., and George W. Bohrnstedt, eds. *Sociological Methodology 1969.* San Francisco, Calif.: Jossey-Bass, 1968.

Bridgman, P. W. *The Logic of Modern Physics.* New York: Macmillan, 1927.

Buckley, Walter. *Sociology and Modern Systems Theory.* Englewood Cliffs, N.J.: Prentice-Hall, 1967.

Burgess, Robert L. "An Experimental and Mathematical Analysis of Group Behavior within Restricted Networks," *Journal of Experimental Social Psychology* (1968), *4:*338–349.

———, and Don Bushell, Jr. *Behavioral Sociology: The Experimental Analysis of Social Process.* New York: Columbia University Press, 1969.

Costner, Herbert L., and Robert K. Leik. "Deductions from 'Axiomatic Theory'," *American Sociological Review* (1964), *29:*819–835.

Edwards, Ward, Harold Lindman, and Leonard J. Savage. "Bayesian Statistical Inference for Psychological Research," *Psychological Review* (1963), *70:*193–242.

Ehrlich, D., I. Guttman, P. Schonbach, and J. Mills. "Postdecision Exposure to Relevant Information," *Journal of Abnormal and Social Psychology* (1957), *54:*98–102.

Festinger, Leon. *A Theory of Cognitive Dissonance.* Stanford, Calif.: Stanford University Press, 1957.

Fisher, Sir Ronald A. *Design of Experiments.* 8th ed. New York: Hafner, 1966.

Freeman, Linton C. *Elementary Applied Statistics: For Students in Behavioral Science.* New York: John Wiley & Sons, 1965.

Hall, Calvin S., and Gardner Lindzey. *Theories of Personality.* New York: John Wiley & Sons, 1957.

Hamblin, Robert L. "Ratio Measurement and Sociological Theory: A Critical Analysis." A report presented at the 1966 Annual Meeting of the American Sociological Association in Miami, Fla. See also description by Robert L. Burgess, "An Experimental and Mathematical Analysis of Group Behavior within Restricted Networks," *Journal of Experimental Social Psychology* (1968), *4:*338–339.

Heider, Fritz. "Attitudes and Cognitive Organization," *Journal of Psychology* (1946), *21:*107–112.

———. *The Psychology of Interpersonal Relations.* New York: John Wiley & Sons, 1958.

Hempel, Carl G. *Fundamentals of Concept Formation in Empirical Science.* Chicago: University of Chicago Press, 1952.

———, and Paul Oppenheim. "Studies in the Logic of Explanation," *Philosophy of Science* (1948), *15:*135–175.

Henicke, Christoph, and Robert F. Bales. "Developmental Trends in the Structure of Small Groups," *Sociometry* (1953), *16:*7–38.

Hilgard, Ernest R., and Gordon H. Bower. *Theories of Learning.* 3rd ed. New York: Appleton-Century-Crofts, 1966.

Homans, George Caspar. *Social Behavior: Its Elementary Forms.* New York: Harcourt, Brace & World, 1961.

Hopkins, Terence K. *The Exercise of Influence in Small Groups.* Totowa, N.J.: The Bedminster Press, 1964.

Kael, Pauline. "The Current Cinema," *The New Yorker* (February 14, 1970), p. 117.

Kasl, Stanislav V., and Sidney Cobb. "Effects of Parental Status Incongruence and Discrepancy on Physical and Mental Health of Adult Offspring," *Journal of Personality and Social Psychology,* (1967), *7* (2), No. 642:1–15.

Kohler, Wolfgang. *Gestalt Psychology.* New York: Liveright, 1929.

Kuhn, Thomas S. *The Structure of Scientific Revolutions.* Chicago, Ill.: University of Chicago Press, 1962.

Lazarsfeld, Paul F., and Neil W. Henry. *Latent Structure Analysis.* Boston: Houghton Mifflin, 1968.

Lewin, Kurt. *Principles of Topological Psychology.* New York: McGraw-Hill, 1936.

Lewis, Donald J. "Partial Reinforcement: A Selective Review of the Literature Since 1950," *Psychological Bulletin* (1960), *57*(1):1–28.

MacCorquodale, Kenneth, and Paul E. Meehl. "On a Distinction between Hypothetical Constructs and Intervening Variables," *Psychological Review* (1948), *55:*95–107.

Michels, Robert. *Political Parties.* Translated by Eden and Cedar Paul. New York: Dover Press, 1959.

National Academy of Sciences—Social Science Research Council (NAS-SSRC). *The Behavioral and Social Sciences: Outlook and Needs.* Englewood Cliffs, N.J.: Prentice-Hall, 1969.

Newcomb, Theodore M. "An Approach to the Study of Communicative Acts," *Psychological Review* (1953), *60:*393–404.

Nisbet, Robert. "Review of Amitai Etzioni, *The Active Society," American Sociological Review* (1968), *33:*988–991.

Osgood, C. E., and P. H. Tannenbaum. "The Principle of Congruity in the Prediction of Attitude Change," *Psychological Review* (1955), *62:*42–55.

Popper, Karl R. *The Poverty of Historicism.* New York: Harper & Row, 1957.

———. *The Logic of Scientific Discovery.* New York: Harper & Row, 1959.

———. *Conjectures and Refutations: The Growth of Scientific Knowledge.* New York: Harper & Row, 1963.

Raiffa, Howard. *Decision Analysis.* Reading, Mass.: Addison-Wesley, 1968.

Rapaport, David. "The Conceptual Model of Psychoanalysis," *Journal of Personality* (1951), *20:*56–81.

Reynolds, Paul Davidson. "Certain Effects of the Expectation to Transmit on Conceptual Attainment," *Journal of Educational Psychology* (1968), *59*(3):139–146.

Rosenberg, Milton J. "Cognitive Structure and Attitudinal Affect," *Journal of Abnormal and Social Psychology* (1956), *53:*367–372.

Rosenthal, Robert. *Experimenter Effects in Behavioral Research.* New York: Appleton-Century-Crofts, 1966.

Secord, Paul F., and Carl W. Backman. *Social Psychology.* New York: McGraw-Hill, 1964.

Siegel, Sidney. *Nonparametric Statistics.* New York: McGraw-Hill, 1956.

Skinner, Burrhus F. "Are Theories of Learning Necessary?" *Psychological Review* (1950), *57*(4):193–216.

Stinchcombe, Arthur L. *Constructing Social Theories.* New York: Harcourt, Brace & World, 1968.

Torgenson, W. *Theory and Method of Scaling.* New York: John Wiley & Sons, 1958.

Webb, Eugene J., Donald T. Campbell, Richard D. Schwartz, and Lee Sechrest. *Unobtrusive Measures: Non-Reactive Research in the Social Sciences.* Chicago: Rand McNally, 1966.

Willer, David, and Murray Webster, Jr. "Theoretical Concepts and Observables," *American Sociological Review* (1970), *35:*748–757.

Author Index

Subject Index*

*Page numbers for definitions are in boldface type.